THE PRIVATE LIFE OF BIRDS

STEPHEN MOSS

First published in 2006 by
New Holland Publishers (UK) Ltd
London • Cape Town • Sydney • Auckland

www.newhollandpublishers.com

Garfield House, 86–88 Edgware Road, London, W2 2EA,
United Kingdom
80 McKenzie Street, Cape Town 8001, South Africa
14 Aquatic Drive, Frenchs Forest, NSW 2086, Australia
218 Lake Road, Northcote, Auckland, New Zealand

10 9 8 7 6 5 4 3 2 1

ISBN 1 84537 422 3
ISBN 978 1 84537 422 8

Editorial Director: Jo Hemmings
Senior Editors: Steffanie Brown and James Parry
Design: Adam Morris and Alan Marshall
Production: Joan Woodroffe

Reproduction by Pica Digital Pte Ltd, Singapore
Printed and bound in Malaysia by Times Offset (M) Sdn Bhd

Pictures appearing on the cover and preliminary pages:
Front cover: clockwise from top: Great Crested Grebe; Black Woodpecker; Wren; White Stork
Spine: Great Egret
Back cover: Sandhill Crane
Opposite: Puffin
Page 2: Blue and Yellow Macaw.
Page 6: *top to bottom*: White Stork; Barn Swallows; Rose-ringed Parakeet
Page 7: *top to bottom*: Willow Warbler; Common Snipe; European Robin

Contents

Introduction 8
How to use this book 10

MOVEMENT 12

Flight 14
Walking, running and climbing 22
Swimming and diving 26
Flocking 32
Roosting and sleeping 36
Moult and feather care 39

MIGRATION 44

What is migration? 46
Migration strategies 53
Unusual migrations 56
Local movements 62
Birds and weather 65
Vagrancy 70

FEEDING 76

What do birds eat? 78
Feeding methods and strategies 80
Predators and prey 90
Specialized feeders 95
Drinking 104

BREEDING 106

The race to reproduce 108
Timing 110
Territory 112
Birdsongs and calls 114
Courtship and displays 125
Nests 128
Eggs 131
Chicks 135
Unusual breeding behaviour 137

WHERE BIRDS LIVE 140

Bird populations 142
Bird distribution 146
Range changes 152
Endangered species 163
Extinction 166

BIRDS AND PEOPLE 168

Birds in history 170
Human exploitation of birds 176
Birds in folklore and culture 183
Conservation and bird protection 188
Birdwarching 191

Glossary 198 • Further reading 199
Useful addresses 201
Index 202 •
Credits and acknowledgements 208

INTRODUCTION

If you have bought, borrowed or been given this book, it is safe to assume that you have at least a passing interest in birds. You may be something of an expert, having spent many years watching or studying them, or a relative beginner, whose interest in birds is quite recent. Either way, I hope that this book has something to offer you; after all, even the most experienced birder has some gaps in their knowledge!

Of course, you may be content to simply watch birds – either in your back garden or local patch, or farther afield in Britain or abroad. But after a while, just *watching* them doesn't seem quite enough; you want to know a bit more about them: their behaviour and movements, distribution and history, even the ways in which they relate to human culture.

That is where *The Private Life of Birds* comes in. This book aims to serve as a basic introduction to different aspects of birds' lives; both the commonplace, day-to-day behaviour, and the more unusual aspects of their existence. These range from 'biological' functions such as courtship, flocking and feeding, through to the 'bigger picture' – to help you put your observations of birds into a broader context – such as distribution, range and population. There is also a chapter on birds and people, which I hope will serve to remind you of the long, and sometimes chequered, relationship between human beings and birds.

The Private Life of Birds can only ever be a brief introduction to the multi-faceted lives of our fellow creatures. After all, whole books have been written about specific types of behaviour, such as those mentioned in the bibliography about bird-song or migration (see page 199). Indeed, if you are fascinated by a particular aspect of bird behaviour you may want to delve even deeper into the life cycle of a particular family or a single species: if so, there are many monographs available.

You may prefer to explore a more cultural aspect of birds' lives; again, I have listed several books covering folklore in the bibliography. And if you want to consult a work of reference to check a particular species or type of behaviour, again there are several such works listed.

If you do not already belong to one or more of the national organizations such as the RSPB, BTO or Wildlife Trusts listed in the Useful Addresses section (see page 201), then it is time you joined at least one of them! There are many benefits for members, including regular magazines and local members groups, or the opportunity to take part in surveys. By joining up you are also giving your support to their valuable and essential conservation work, and helping to safeguard the future of our wildlife.

Finally, I hope that whatever your level of knowledge or experience of birds, you will find this book an enjoyable and informative read, and that having read it you are spurred on to spend more time out in the field, finding, watching and enjoying birds. We have been watching birds – in one way or another – since man first evolved, and enjoying the pastime of bird-watching for at least a century. Today, birds are in trouble as never before, and so they need our help. But we don't watch birds entirely for their own good. We also gain huge benefits from doing so. In the past decade or so we have seen a major shift in our relationship with birds and other wildlife; especially with the creatures and places near where we live. The idea that going out and watching birds is beneficial for one's physical health has always been accepted – even if the main advantage has been getting some fresh air into our lungs. Today, we are increasingly aware of the spiritual and emotional benefits as well. Put simply, watching birds is good for us!

BELOW: Emperor Penguins raise their young in the Antarctic winter, in some of the harshest conditions imaginable.

How to use this book

The book is divided into six chapters, each covering a particular aspect of birds and their behaviour.

Movement
Includes sections on flight (and flightlessness); walking, running and climbing; swimming and diving; and associated behaviour such as flocking, roosting and sleeping, moult and feather care.

Migration
Includes sections on the basics of migratory behaviour; migration strategies; unusual forms of migration; local movements; birds and weather; and vagrancy.

Feeding
Includes sections on choice and types of food; feeding strategies; unusual feeding behaviour; predators and prey; specialized feeders; and drinking methods.

Breeding
Includes sections on general reproductive-behaviour; territory; birdsongs and calls; courtship displays; nests; eggs; chicks; and unusual breeding behaviour.

Where Birds Live
Includes sections on bird populations; bird distribution and range; range changes; endangered species; and extinct species.

Birds and People
Includes sections on the historical relationship between birds and people; the human use of birds for domestication and hunting; cagebirds; birds in high and popular culture; conservation and bird protection; and birdwatching.

These chapters are further sub-divided into sections for ease of use, with short essays about particular aspects of the lives of birds. You can, if you wish, read the book from cover to cover, although you may find it more useful to dip into, or to look up a specific aspect of behaviour in the index. I have tried to write for a general audience, assuming no prior knowledge; though inevitably I have occasionally had to use specialized terms. If you are unsure what one of these means, there is a glossary at the back of the book to help you. I have also included a short section of useful addresses (conservation organizations such as the RSPB, plus some commercial companies to enable you to obtain books and bird food) and a list of books covering a range of subjects. If your appetite for knowledge has been whetted by reading this book, these works provide more detailed and in-depth coverage of the subjects included here.

OPPOSITE TOP LEFT: A Kingfisher with its typical prey of small fish.
OPPOSITE TOP RIGHT: A Great Grey Owl with young at the nest.
OPPOSITE: The Evening Grosbeak has one of the most powerful bills of any songbird.

MOVEMENT

The way birds move is often the first thing we notice about them. Whether flying through the air, swimming across the surface of a lake or pond, or walking or hopping across a lawn, the method chosen is a vital clue to a bird's identity and general behaviour.

Flight is the mode of travelling we most associate with birds, and all but a tiny minority of the world's ten thousand or so species can fly. But for the six hundred or so species classified as 'waterbirds', swimming and diving are frequently just as important. Whilst other species will occasionally take to the water, many – including more than half of the world's species, the passerines or perching birds – prefer a more terrestrial form of travel.

This chapter covers all areas of locomotion: flight (and flightlessness), walking, running and climbing, swimming and diving, along with associated behaviour such as flocking, roosting and sleeping, and moult and feather care.

RIGHT: Hummingbirds are amongst the world's most proficient flyers and able to hover in mid-air in order to feed.

Flight

To many of us, the very thing that defines birds is their ability to fly. All but fifty or so of the world's ten thousand species of bird can fly, and even those that do not, such as the Ostrich, Kiwi and Emu, evolved from birds that did once take to the air. While insects and bats are also pretty good flyers, birds are arguably the greatest masters of the air, having evolved so many different ways of getting and staying aloft.

HOW BIRDS FLY

Feathers are the key to birds' ability to defy gravity: they are light, strong and supremely manoeuvrable, enabling their owners to glide, soar and hover where necessary. Birds also have a specially adapted skeleton, with hollow bones to reduce body weight; these are reinforced with a series of struts to provide strength and flexibility: both features are absolutely vital, especially as the bird takes to the air.

Birds' wing feathers follow a set pattern, with the longest primary feathers used to provide lift, while the secondaries and tertials enable the bird to stay aloft. Together they provide a superb solution to the various problems of flight.

Soaring success

Once airborne, birds use a variety of

LEFT: Large and powerful raptors such as the Bald Eagle fly by soaring on broad wings.

ABOVE: The Barn Owl hunts on silent wings, their white appearance giving the bird a ghostly quality.

techniques to stay there. Gliding and soaring enable them to use as little energy as possible while remaining aloft, while more refined techniques – such as hovering motionless in one position – enable birds such as Kestrels and hummingbirds to find food.

Birds will adjust their wing shape in order to suit the prevailing conditions, and also to adopt the most appropriate flight style. Thus, a bird of prey may open its wings fully when soaring, and then close its wings to make its body shape more aerodynamic in order to dive down onto its unsuspecting prey.

Long haul flights

Migration puts a lot of strain on birds: they must minimize their use of energy while maximizing the distance flown. Some do so by travelling at night, when flying is easier because cool air provides less wind resistance. Others travel in groups, rather like racing cyclists, with the leader taking the strain while those following behind benefit from travelling in the slipstream.

For large birds such as pelicans, storks and eagles, travelling long distances is made even harder by the difficulty they have in getting and staying airborne. Such groups take advantage of thermals – rising air currents, usually found at the edge of

ABOVE: Like many large birds, ibises will migrate in a 'V'-formation to save energy when flying long distances.

OPPOSITE: The Common Swift spends almost its whole life on the wing, only landing during the breeding season.

mountains and other raised areas. Large birds find it especially difficult to fly across large open areas of water, simply because thermals either do not occur or are much weaker there.

Seabirds are perhaps the greatest masters of the air: gliding for hours or even days on end with hardly a flap, using a technique known as dynamic soaring.

High flyers

The question 'how high do birds fly?' is not easy to answer: like many aspects of bird behaviour, it depends on the bird in question and the circumstances in which it finds itself. So most songbirds spend their lives flying more or less at ground level, but on migration they usually travel at a height of about 1500 feet (450m). Some birds of prey will migrate at higher levels, perhaps to 3,300 feet (1000m), while some waders and wildfowl may rise even higher. The record altitude is that of a Rüppell's Griffon Vulture, sucked into the jet engine of an aircraft over West Africa at an astonishing 37,000 feet (11,300m) above sea level – higher than Mount Everest! The highest altitude at which a flying bird has actually been observed was achieved by a flock of Bar-headed Geese, seen above Everest at about 31,000 feet (9,500m) above sea level.

THE KINGS OF THE AIR

When it comes to flying, the champions – at least when it comes to endurance – fall into two groups: seabirds and swifts. Both manage to stay aloft for incredible amounts of time (possibly as long as ten years), and do so as complete masters of the skies, making the very act of flying look supremely easy.

Cruising at sea

Among seabirds, the most stylish of the flyers are the shearwaters and albatrosses. Both share a similar structure: long, narrow wings, tapering to a point, and held stiffly to the sides in order to maximize the airflow to gain lift, while minimizing drag caused by air resistance. Like beautifully designed aircraft, they seem to belong in the air; unlike aircraft, they are much more flexible and manoeuvrable, and can fly barely above the surface of the sea. Their technique is a simple one: updraughts from the surface of the ocean provide lift, and like a fixed wing aircraft, the difference in air pressure above and beneath the wing provides lift and forward motion.

Using this simple but highly effective technique, these magnificent creatures can cruise around the world's oceans for weeks or even months on end, in search of food for themselves and their young. Many follow ocean-going ships, which provide an added advantage in that this reduces the wind resistance, enabling the bird to expend even less energy.

Other types of seabird use different techniques to stay aloft. For example, frigatebirds and certain species of tern do so by maximizing their wing area in relation to their body weight, thus reducing what aero-engineers call 'wing-loading'. The greater the wing area and the lower the bird's weight, the easier it is to stay aloft. Frigatebirds in particular can simply hang in the air, without having to flap their enormous wings. Sooty Terns are similar in shape, and are reputed to spend as long as seven years in the air before returning to land to breed.

Aerial display

In the UK, the most aerial bird is undoubtedly the Common Swift. Swifts return from their African wintering grounds in late April or early May, and can be seen flying low around the roofs of towns and cities, screaming as they go. They do land to breed, but only perching in the eaves of buildings where they build their nests. After hatching, however, a Swift will leave the nest and will not touch down again for 21 months or more, having travelled back and forth to Africa twice in the meantime!

Swifts are perfectly adapted to an aerial existence: they have long, slender wings and incredibly light bodies. Not only do they feed on the wing; they even sleep there, rising high into the sky and snatching brief periods of slumber for a few seconds at a time.

COPING WITHOUT FLIGHT

Only around 50 of the world's bird species (around 0.5 per cent of the total) cannot fly, having lost the power of flight sometime in their distant evolutionary past. Given the huge success of the world's flying birds, it might be assumed that giving up the ability to become airborne would be a risky strategy. In fact, it can work very well; just so long as the environment in which a flightless bird lives remains similar to the one in which it originally evolved.

Heavyweights

Take the Ostrich. A finer example of a bird adapted to survive the dangers of the African savannah would be hard to find. While it can no longer take to the air to escape attack by predators such as big cats, by giving up the ability to fly, it has been able to grow far bigger and heavier than any bird that retains the power of flight. Ostriches weigh well over 280 lbs (130kg), at least seven times the weight of the heaviest flying bird. As a result, they are rarely killed by predators, and if attacked, usually respond with a hefty and effective kick from their powerful legs and feet.

Other continents have their equivalents of the Ostrich: the rheas of South America, and the Emu and Cassowary of Australasia. All share the same body shape: feathers that have become more like mammalian fur; huge eyes and long necks (to enable them to look out for predators), and of course, their large size. But one branch of this group – known as ratites – has evolved in rather different ways: the kiwis of New Zealand.

BELOW: The world's heaviest bird, the Ostrich may be unable to fly but it can run extremely fast!

OPPOSITE: The Dodo's flightlessness made it highly vulnerable to imported predators such as cats and rats.

Flightlessness and extinction

The benefits of flightlessness are clear, but what about the drawbacks? Until human beings began to explore the world's distant regions, most flightless birds were successful, living on remote oceanic islands, whose impoverished variety of creatures meant that there were no ground-dwelling mammalian predators.

As a result, many island birds became flightless: especially rails, some parrots, and the Dodo and the Solitaire, relatives of the pigeons and found on islands in the Indian Ocean. Up to five hundred years ago, these bizarre creatures thrived, at least within their limited ranges. Then came the Age of Exploration, with fleets from England, France, Spain and Portugal ranging across the globe. These ships carried hungry sailors, who found the presence of the ludicrously tame and trusting birds irresistible; indeed, the first fresh meat they had tasted for months. So it was that the plump and apparently tasty Dodo met its inevitable fate: a century or so after the first explorers arrived on Mauritius in the 1500s, it was extinct.

Other less edible and more obscure flightless species lasted a little longer. But their fate was sealed as soon as the explorers came ashore. For along with them they brought dogs, cats and pigs, all of which were able to take advantage of flightless birds. Worst of all were rats, which escaped from the ships' holds, bred rapidly, and wreaked havoc on the populations of all ground-nesting birds, whether flightless or not.

Small benefits

The four species of kiwi – all confined to New Zealand, where they are the national bird – share many features of their relatives, but have one major difference: size. Whereas Ostriches and rheas grew bigger in order to survive, kiwis grew smaller, to exploit the absence of insectivorous mammals in New Zealand. We know that kiwis evolved from much larger birds for one simple reason: their eggs are huge, comprising up to one quarter of the female kiwi's body weight (compared with about 1.5 per cent for the Ostrich). This tells us that although the birds evolved to become smaller because of the advantages conferred, there was not as much evolutionary pressure to reduce the size of the eggs. Hence, the female kiwi has one of the greatest challenges faced by any mother!

WEATHER AND FLYING BIRDS

Flying birds are arguably more affected by weather conditions than any other living creature: not only do many of them have to find food in the air, but birds undertake vast global journeys of migration, encountering all sorts of different weather conditions, from wind to rain and thunderstorms to droughts.

Locally, weather can affect birds in all sorts of ways: a particularly heavy rainstorm or gales may force them to land, and in extreme cases, birds have been killed by gales or hailstorms. But it is on their epic migratory travels that birds are particularly affected. Being blown off their intended course is probably the greatest hazard: either by normal winds (a process known as drift) which particularly disorients young birds on their first journey south; or by stronger gales, which affect all birds, forcing them to make landfall or risk falling into the sea and drowning.

BELOW: In fine weather birds such as Sandhill Cranes can travel vast distances.

An ill wind

During easterly gales across the North Sea, especially when driving rain reduces visibility, birdwatchers on the east coast of Britain watch out for 'falls' of warblers, flycatchers and chats, especially on islands or headlands, where they concentrate sometimes in their thousands. As the weather improves, the birds will feed, replenish their energy resources, and then carry on, resuming their intended course. If the weather does not improve, however, many will die of starvation, or will become so disoriented that they actually get lost – what birdwatchers call 'vagrants'.

Vagrancy does not only occur in 'bad' weather: the phenomenon known as 'spring overshooting' occurs when high pressure systems bring fine, settled weather to Europe in spring. As birds return to breed in southern Europe, many simply keep on going, ending up much farther north than they intended. In good years, large numbers of species such as egrets, herons, European Bee-eaters and Hoopoes turn up in southern England.

Walking, running and climbing

Unlike mammals, which generally run (apart from kangaroos, wallabies and odd-balls such as the lemurs of Madagascar), birds choose a variety of ways of getting around on the ground. These include hopping, walking and running. Several different, unrelated groups of birds have evolved into superb climbers: usually of trees, but also of rocky cliffs.

HOP, WALK OR RUN?

Hopping is practised by the majority of passerines, the category which makes up more than half the world's birds. Thrushes, sparrows, finches, buntings and robins hop; while crows, wagtails, pipits and starlings walk. Among non-passerines such as waders, gulls and wildfowl, walking is the norm. Sometimes this turns into a short run; especially when a morsel of food is at stake, and another bird might reach it first. But by and large, birds rarely run for more than a short distance.

BELOW: Roadrunners spend most of their time on the ground and chase down their prey on foot.

Speed merchants

There are exceptions to this rule, of course. Sanderling gather in little groups along the tideline, waiting for the waves to recede, dashing forward to grab a morsel of food, then running back up the sandy beach to avoid getting their feet too wet. Ostriches and rheas can turn in an incredible burst of speed, especially when they think that their young might be in danger. Those long, muscular legs are perfect for sprinting, and even the babies manage to keep up with their parent, at least over a

OPPOSITE: Woodpeckers are amongst the best adapted of all the world's birds for climbing trees.

short distance until they are away from imminent danger. Ostriches have the title of the world's fastest bird on land, and are able to reach speeds of more than 40 mph in short bursts; while even at normal pace they run at about 30 mph.

Front runner

But the greatest runners of all are the birds whose very name sums up their skill: the two species of roadrunner. The Greater and Lesser Roadrunners are essentially ground-dwelling cuckoos. They have adapted to a terrestrial existence (though they retain the powers of flight) by their long legs and large tail, which they use as a rudder when reaching speeds of up to 25 mph. Found in the deserts of the New World, from the southern states of the USA through Mexico to Central America, they use their ability to run to chase and catch a variety of prey items, including scorpions, snakes and lizards – all creatures that usually rely on speed in order to escape. Roadrunners achieved fame through the popularity of the American cartoon series featuring Wile E. Coyote and his adversary (who always seemed to gain the upper hand in their encounters!), the eponymous Roadrunner. Beep-beep!

CLIMBING – UP AND DOWN!

The group of birds most supremely adapted to an arboreal life is surely the woodpecker family, whose 215 or so species are all perfectly adapted to climbing. One of the oldest forms of birdlife, woodpeckers split from their closest relatives at least 50 million years ago, giving them plenty of time to develop their unique adaptations for climbing.

Woodpeckers are adapted to their lifestyle in a number of ways. All have strong claws, usually with four toes (though some with only three), which enable them to grip onto the vertical surface of a tree

(and indeed, to move around a branch, often clinging on from beneath and further defying the pull of gravity). They also have a long tail, with specially strengthened central feathers, which enable them to use it as a prop in order to further strengthen their ability to cling on. Indeed, many woodpeckers moult these central tail feathers last of all, to retain their climbing abilities for the greatest possible time.

Jeepers creepers

Not all woodpeckers live on trees, however. Apart from the more terrestrial species (such as the Ground Woodpecker of South Africa, or the flickers of the New World), other species survive on large cactus plants, and so are able to live in desert areas without trees. However, the vast majority do spend their lives in woods and forests, often moving only short distances in what is one of the most sedentary lifestyles of all birds.

Two other closely related groups of birds are also able to climb, and have similarly sedentary lifestyles. Treecreepers live up to their name by climbing around the trunks and branches of trees, hugging the bark, and like woodpeckers, using their tail and sharp claws to stay on. Nuthatches are unique: the world's two dozen or so species are the only birds in the world capable of climbing down as well as up! Finally, the Wallcreeper, a stunningly beautiful relative of the rather dowdy and inconspicuous treecreepers, lives on steep crags and cliff faces in mountainous areas of Europe, Africa and Asia.

LEFT: The Treecreeper is superbly adapted for an arboreal existence on the trunks and branches of trees.

OPPOSITE: All types of jacana are able to walk on surface vegetation, earning them the nickname 'lily-trotter'.

Walking on water

Jesus walking on the waters of Lake Galilee may have been a miracle, but for one group of birds, walking on water is part of their daily lives. The eight species of jacana are found in tropical regions of South and Central America, sub-Saharan Africa, the Indian sub-continent, South-East Asia and northern Australasia, and all share one extraordinary ability: that of appearing to walk across the surface of lakes, ponds and marshes.

Of course the jacanas cannot actually walk on water; but their extraordinarily long toes do enable them to walk across submerged aquatic vegetation, such as water-lilies, without sinking beneath the surface. Their large 'footprint' means that their weight is spread out, enabling them to venture where other birds would fear to tread.

Jacanas look like large, thin rails, but in fact they are more closely related to waders. Most are not particularly good flyers, though they will move between areas of water, especially during times of drought when their original home may dry up. They nest on the surface of the vegetation, piling up a small amount of plant material and laying their eggs on top. As a result, the eggs sometimes sink into the water, and may even become lost.

Because of this amazing ability to walk across bodies of water using vegetation as a platform, jacanas have earned the alternative name of 'lily-trotter'.

Swimming and diving

About four hundred bird species regularly swim: notably the 160 or so species of wildfowl (ducks, geese and swans), but also grebes, divers, pelicans, cormorants and shags, some species of rail (the coots and gallinules), and a number of seabirds, including albatrosses, shearwaters, gulls, terns, auks and a strange family of waders, the phalaropes.

IN THE SWIM

In order to swim effectively, all of these species have some degree of webbing between their toes, which helps give effective propulsion when moving through the water. The amount of webbing in species varies considerably, from full webs (ducks, geese and swans) to partially webbed or lobed feet (coots, phalaropes and grebes).

Other adaptations to an essentially aquatic life include the ability to waterproof feathers, especially vital for those birds that spend the majority of their lives on or under the water, such as ducks. Birds keep their feathers waterproof by spreading an oily secretion from a gland at the base of their tail, using their bills to coat each feather tract.

BELOW: Webbed feet are essential for a successful aquatic existence.

OPPOSITE: The penguin's streamlined shape is ideal for swimming underwater.

Champion divers

Of all the world's birds, the champion divers are a group which has abandoned the ability to fly in order to adapt to an essentially aquatic existence: the penguins. The world's 17 or so species of penguin are confined to the southern hemisphere, although despite their popular image, not all live around the Antarctic.

All are brilliantly adapted to their undersea lifestyle. Physically, penguins have evolved to function underwater rather than on land: their wings have lost their ability to support flight, instead becoming powerful flippers which propel the birds at speeds much faster than any human swimmer can achieve.

They can also reduce their heartbeat when diving, from a typical rate while resting on land of about 80–100 beats per minute (a little over the human average) to a mere 20 beats per minute when underwater. This reduction helps eke out their oxygen supply, and means that penguins can habitually remain underwater for five or six minutes at a time – a huge advantage when searching for food. The record dive was timed at 18 minutes, carried out by an Emperor Penguin, the largest member of the family.

Penguins can also reach great depths: more than 1,600 feet (500m) below the surface, again achieved by an Emperor Penguin. More typically, they tend to dive to about 160–330 feet (50–100m) beneath the surface.

Feet last

Buoyancy can be both useful and a hindrance: birds sitting on the surface of the water want to float easily, but when the same bird needs to dive deeply, it has to fight its natural tendency to float. So many diving birds, especially divers and grebes, have feet well to the back of their body, which act as powerful tools to propel the bird down. They also streamline their body in order to reduce water resistance. As a result, however, divers and grebes in particular really struggle when they come onto land, as what works in the water is a handicap when trying to walk!

DIPPERS – THE SUBMARINE BIRD

We do not usually associate passerines – the order that contains well over half the world's bird species – with water. And it is true that of almost six thousand species, only five habitually submerge themselves as part of their normal daily lives: the members of the dipper family.

Dippers are unique amongst their fellow passerines in that they are the only species which feed beneath the surface of the water. They live alongside fast-flowing streams and rivers, and frequently plunge right into the rapidly moving waters in order to grab small morsels of food, mainly aquatic insects and other small invertebrates. However, their name derives not from this unique adaptation, but from their characteristic habit of bobbing up and down when looking for food – perhaps a distraction technique designed to break up their body outline, making them less conspicuous to their prey.

There are five species of dipper in the world, found on all the world's continents apart from Australasia, and of course, Antarctica (though with only a toehold in North Africa). The two most familiar species, the White-throated or Eurasian Dipper and the American Dipper, are well studied, and have provided many insights into how this amazing bird became adapted to its unusual aquatic lifestyle.

OPPOSITE: The smallest member of its family in Britain, the Little Grebe is also known as the Dabchick on account of its cute, fluffy appearance.

BELOW: Red-throated Divers appear clumsy on land, as their bodies are better adapted to an aquatic existence.

Walking underwater

The dippers have evolved several specific body features that enable them to dive and walk underwater. First, their body shape: stocky yet streamlined, with a thick covering of feathers around the plump little body, which helps retain heat and avoid rapid cooling when wet. Dippers also have short, but powerful wings – rather like auks or penguins – which act more like flippers, though they can also be used for rapid flight. However, dippers are not designed for migration: most lead very sedentary lives along the same stretch of river or stream, though those living in mountainous areas of Europe, Asia or North America may undertake altitudinal movements between summer and winter, moving several thousand feet lower in cold weather.

Like more traditional waterbirds such as

Grebes – the most aquatic bird

Ask most birdwatchers which group of birds lives the most aquatic lifestyle and the chances are that they would suggest ducks, or perhaps one of the ocean-going seabird families such as albatrosses, petrels or shearwaters. Yet the real candidates for this title are the world's 22 or so species of grebe.

Grebes feed and breed on water, building their nests and raising their young there too. The nest is a floating structure among the aquatic vegetation near the edge of a pond or lake. Adults rarely emerge onto land at all, and their babies even ride on their backs when they need a break from swimming!

While looking superficially like ducks, grebes are in fact more similar in form and structure to coots. Like coots, they have straight, narrow bills (unlike ducks) and lobed feet instead of webbed ones. They come in several shapes and sizes, ranging from the tiny Least Grebe of the Americas to the elegant Western and Clark's Grebes of the USA, which are about four times the length of their smallest relative and more than sixteen times the weight!

Grebes dive for food, and their slender, streamlined bodies and numerous tiny feathers make them highly adept at doing so. Diving is also an integral part of courtship, especially for the Great Crested Grebe, with both male and female diving for water-weed which they then offer to each other in an extraordinary ritualized dance (see page 126).

ABOVE: Although songbirds, dippers are superbly adapted to life in, on and under the water .

OPPOSITE: The Grey (or Red) Phalarope is one of only three species of wader that habitually swim.

ducks, dippers have a specially developed preen gland at the base of the tail, which secretes oil which the bird can then use to preen its feathers and keep them waterproof. This is especially important given that the bird needs to submerge very frequently when feeding, usually staying under for between ten and thirty seconds before emerging.

Keep it clean

For all species of dippers, clean, clear and pollution-free water is essential. There have been fears that dippers in Europe might decline because of the onset of 'acid rain', emissions from power stations that fall as rain and enter the water supply. Acidification of highland rivers because of the planting of conifers is another problem. Fortunately, such fears do not yet seem justified, though the Eurasian Dipper has undergone a decline in population in Britain in recent years.

PHALAROPES – THE WADER THAT SWIMS

Most waders wade for a living, which is how they got their name. Some, especially long-legged species such as godwits, avocets and shanks, will also swim occasionally, usually when they have waded further into the water than usual. But the members of one family, comprising just three of more than two hundred wader species habitually swim: these are the phalaropes.

All three species of phalarope breed in what biologists call the 'Holarctic' region: the part of the northern hemisphere that comprises the whole of the Arctic, including the northern regions of North America, Europe and Asia. All three species are also long-distance migrants, heading on epic journeys south each autumn to winter in the tropical oceans or even right down into the southern hemisphere. Like grebes and coots, two of the three species (Red-necked and Grey or Red) have toes with tiny flaps of skin known as lobes, which help them to propel themselves through the water.

In a spin

Phalaropes are tiny birds: the smallest species (Red-necked) being little bigger than a sparrow in length and weight. Given their small size, it seems extraordinary that they can live in the open oceans, but that is just what they do outside the brief northern hemisphere breeding season. In fact, they are supremely adapted to such an aquatic lifestyle, being highly buoyant because of their ability to trap air beneath their feathers.

During the breeding season itself, phalaropes live on small bodies of water such as marshes and ponds, or in shallow coastal waters, where they feed by picking tiny morsels off the surface with their needle-shaped, pointed bills. They also use their skill as swimmers to 'spin', manoeuvring themselves in tight little circles in order to stir up small aquatic creatures from the water and mud below.

Flocking

One of the most noticeable characteristics of birds is their sociability. Although not all species do so, many birds regularly come together with others of their kind, in what we know as a flock.

Birds flock for several reasons, mostly connected with three main types of behaviour: feeding, safety and security, and migration. Thus a particular species may be solitary, or may live in pairs for much of its lifecycle, but it will come together with others at a particular time of day, season or year.

WHY BIRDS FLOCK

Flocking to find food is the most frequent reason why birds will form flocks. Flocking may occur spontaneously or regularly, depending on the type of bird involved and the kind of food it eats. Thus a glut of fish off the end of a trawler will inevitably attract a large and mixed flock of seabirds; mainly gulls, but also including gannets, shearwaters, petrels, skuas and terns, depending on the location and season. Similarly, the death of a large animal in the African savannah will rapidly attract a feeding flock: this time mainly comprising vultures, Marabou Storks and other scavengers. And putting out food for birds in a

BELOW: Red-billed Queleas, the world's commonest bird, gather in huge flocks to feed and drink.

garden can also attract flocks, such as finches and Starlings.

Other food-related flocks occur at certain times of year, such as the mixed flocks of tits that come together during the autumn and winter to forage for food in woods, hedgerows and gardens. These keep together using what are known as 'contact calls' – simple notes that enable each bird to stay in touch with the main group. Such flocks may consist of several different species of tit, together with the odd Nuthatch, Treecreeper and Goldcrest, coming along for the ride. They join together because food is scarce at this time of year and more eyes searching are more likely to find it.

Safety in numbers

The second reason for flocking – safety and security – is often closely related to the finding of food. After all small birds can easily be distracted when feeding or drinking, and thus may be easy prey for a watchful Sparrowhawk or some other predator. By joining together and increasing the pairs of eyes looking out for danger, birds such as tits enjoy the safety of being in a group. If one does spot a predator it will immediately sound the alarm and then flee, giving its fellow birds a sporting chance of escaping alongside it.

Birds that depend on low tide in order to feed, such as waders, are also prone to flocking. Because the time of high and low tide changes from day to day, and because they sleep at high tide, these birds run the risk of attack from a range of predators, from the day-flying Peregrine to the night-hunting Fox. Again, by joining with others in a flock, each individual minimizes the risk to itself.

Flocks of pests

Many farmers dread the sight of large feeding flocks, knowing that they can destroy a crop (by feeding on it or trampling it) in days – sometimes even hours.

In Britain, the biggest problems of feeding flocks are caused by wild geese during the winter months. Species such as Barnacle, Pink-footed and White-fronted Geese like nothing better than to feed on a crop of potatoes, sugar beet or corn, and can cause a great deal of damage by doing so.

But the destruction caused by geese is nothing compared to the amount of damage that one particular species can do in Africa. The Red-billed Quelea is a member of the weaver family, related to our sparrows, and at just a few centimetres long it does not look as if it could do much harm. But when these little birds gather in flocks numbering as many as 100 million individuals, then they become a serious agricultural pest.

It is said that a flock of queleas can strip a field bare in a matter of minutes; and in a region where every ounce of food is precious, they represent a serious problem. Various measures have been tried to get rid of them, including spraying poison from the air and setting fire to their nesting colonies. Nothing seems to work, however, and despite its comparatively restricted range (sub-Saharan Africa), the Red-billed Quelea is the world's most common bird.

ABOVE: Brent Geese are one of several goose species that take advantage of farm crops for autumn and winter food.

On the move

The final reason for flocking is practised by some – although by no means all – migratory birds. To find out why, watch the Tour de France bicycle race, and you will soon see how riding in a pack helps reduce wind resistance for all those cyclists following the lead rider. From time to time, riders will swap positions in the *peloton*, taking turns at the front for a time before dropping back to rest again. The same is true of birds that migrate in a 'V'-formation, notably geese and cranes.

Migrant birds that are normally solitary, such as large birds of prey, will also flock during spring and autumn migration, though more from necessity than desire. This is because they are unable to fly across large expenses of open water, so they must cross from one landmass to another by the shortest possible route. For this reason, hundreds of birdwatchers gather during busy migration periods at places such as Gibraltar, the Bosphorus (at Istanbul), and Eilat, Israel, to enjoy the spectacle of huge flocks made up of millions of migrating raptors.

Roosting and sleeping

Humans usually have a regular sleep pattern consisting of a single major period of sleep, about six to nine hours long, at night. Birds' sleeping patterns are far less regular and more complex. Most diurnal birds sleep for a single extended period each night, usually beginning at dusk and ending at dawn. But many other birds sleep in far less predictable ways.

SLEEPING PATTERNS

Birds that depend on the motions of the tide to feed, such as waders and several species of wildfowl and heron, adopt a very different rhythm from those that do not. Thus, as the tide begins to rise, they will gather together in small flocks, eventually joining together with others into a large flock, often at a single high point known as a roost. There, they will sleep (though often only for a succession of brief, snatched periods) until the tide begins to drop. Gradually, as each new area of mudflat begins to emerge, groups of birds will wake up, fly off and resume feeding. The same cycle is repeated, with minor variations, twice each day.

Where birds sleep also varies consider-

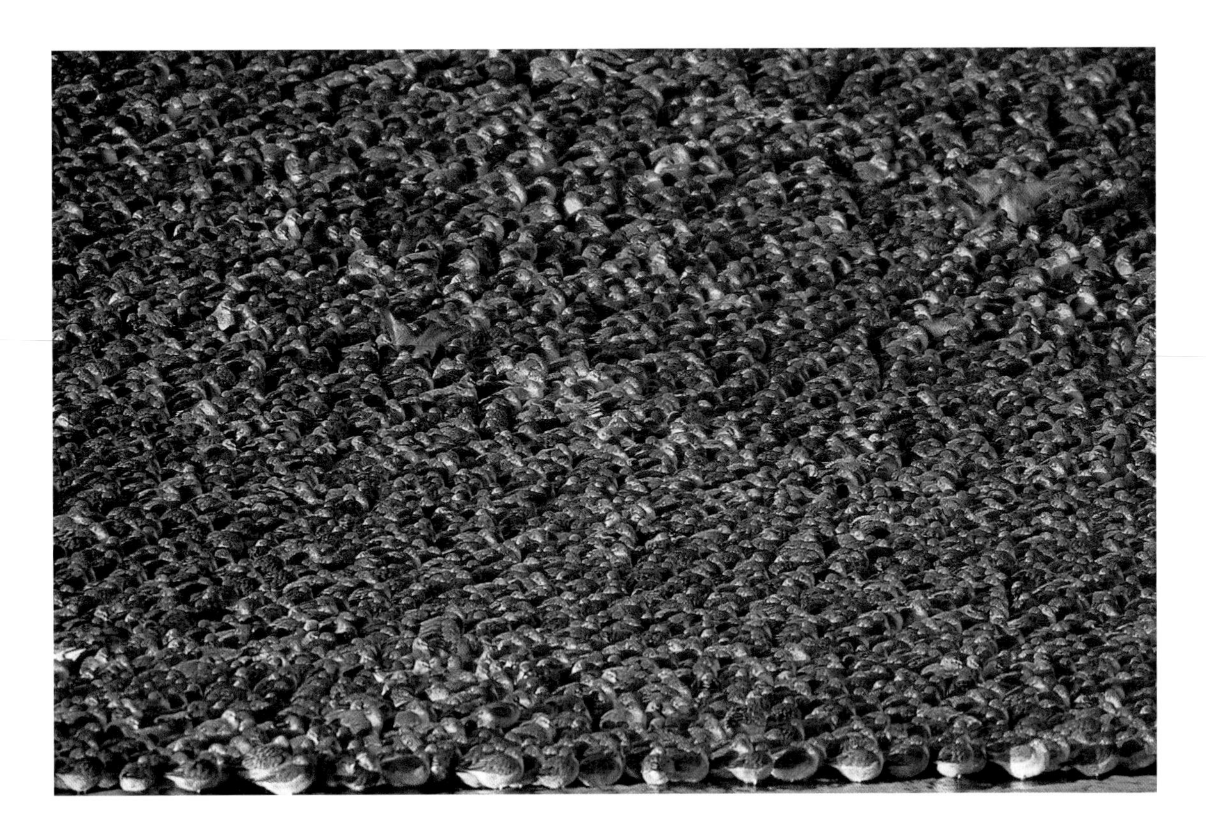

OPPOSITE: Many species of owl, such as this Tawny, are active only at night. They sleep by day and can therefore be difficult to find.

ABOVE: Many waders, such as these Knot, gather to roost in vast numbers when high tides mean that they are unable to feed.

ably from species to species, and even from season to season. And so during the breeding season, birds such as Starlings and Pied Wagtails will stay in pairs or roost in small groups; but as cold weather arrives they tend to form larger roosts, in some cases travelling many miles in order to do so. These roosts are often in the centre of towns or cities, where the birds can take advantage both of the increased warmth from the 'urban heat island' effect and the relative lack of predators, put off by bright street lights.

STARLING ROOSTS

Anyone who was born before about 1970 will recall, at some time in their childhood, becoming aware of large numbers of birds gathering at dusk in a local town or city centre, or perhaps in the open countryside. These were Starlings, and these gatherings were predominantly a winter phenomenon in which tens of thousands – sometimes even millions – of individual birds would join together to form a single, noisy roost.

Today there are hardly any of these huge and spectacular roosts left. Those in London's Leicester Square, or Bristol's Temple Meads station, for example, are long gone. Others in agricultural areas such as Kent and Sussex also appear to have declined or disappeared for good – though that at Brighton's famous pier is still attracting sightseers on winter evenings.

ABOVE: *Starlings massing over the pier in Brighton, Sussex. This is one of the largest roosts for this species in Britain.*

Disappearing act

The reasons for this phenomenon are not easy to explain. Starlings in particular have experienced a major population decline; while it is true that some roosts were driven away by owners of buildings, fed up that their stonework was being damaged by the cumulative effect of tons of droppings, it seems unlikely that so many roosts have disappeared simply as a result of this. Another, more plausible, reason is that milder winters on the European continent have meant that the millions of Starlings that used to travel west to spend the winter in Britain no longer do so.

There are still some roosts in the UK, including a large one on the Somerset Levels. This roost, which attracts several million birds in mid-winter, is unusual in that the birds drop into reedbeds at dusk; a way to avoid predators such as foxes.

OPPOSITE: Gulls such as this Lesser Black-backed moult for several years before attaining adult plumage.

How do penguins keep warm?

Not all of the world's 17 species of penguin live in the Antarctic, but for those that do, the greatest challenge is keeping warm. Some achieve this through their anatomy and physiology, and others by changes in their habits.

The feathers of a penguin are adapted to life in cold oceans: they are more like mammalian fur than that of any other bird, being short, dense and perfect for trapping a layer of warmer air to insulate their bodies against the cold.

Penguins are experts at regulating their body temperature, too. This doesn't just apply to those species that live in the Antarctic region; indeed, regulation is a problem for species such as the Galapagos Penguin, which lives on the Equator but hunts for food in icy cold seas.

But the greatest challenge of all is surely that faced by Emperor Penguins, which spend the entire Antarctic winter standing on the icy ground, brooding their single egg. They have evolved a splendidly cooperative means of keeping warm, huddling together in a dense flock to reduce the effects of the biting wind.

Moult and feather care

Moult is a vital part of a bird's lifecycle, for if it did not shed its worn feathers and grow spanking new ones, it would eventually find it harder and harder to live a normal life, and especially to fly and migrate. Plumage must be kept clean if it is to provide essential warmth and waterproofing; it is also harder to fly with dirty feathers. Regular bathing and preening sessions are essential.

MOULT

If one were to visit an ancient, broad-leaved woodland in Britain in May, it would be alive with the sound of birdsong and the sight of males and females of various species going about their business in the race to reproduce. Go to the same woodland in winter, and it may take a little longer to find the birds, but if you listen for the thin, high-pitched calls of a tit flock, you will soon locate them and see them busily feeding to stoke up energy for the cold winter night ahead.

But go to the same wood on a warm summer's afternoon, say in July or early August, and you could be forgiven for assuming that there are no birds there at all. Not only are there no birdsongs or calls, but despite searching and scanning, you are unlikely to see more than one or two solitary individuals.

Yet there may be more birds in the wood at that very time than at any other season of the year. The difference is that in high summer, most songbirds lose their feathers and regrow new ones, a process known as moulting. At this time of year, these birds are highly vulnerable to predators, being less able to fly properly and escape attack. And so they lie low – or in this case, high – staying hidden in the dense canopy, out of sight.

Eclipse in ducks

We all know about eclipses of the sun and moon, but the word is also used by ornithologists and birders to refer to a very specific form of moult, found mainly in members of the duck family.

Technically known as 'post-nuptial moult', it occurs immediately after the breeding season is over. The males of some species, such as Eider, don't even wait that long; as soon as the chicks have hatched they are off to the local rocks to shed their splendid black and white plumage, while the females do all the hard work guarding the ducklings in a crèche.

Many ducks choose to lie low during their eclipse period, and for very good reason. Because they lose all their flight feathers at once, they are incapable of flying – and therefore achieving a rapid escape from predators such as foxes – for as long as a couple of weeks to a month.

Eclipse plumages are highly confusing to more advanced birders as well as complete beginners. Beginners tend to wonder where all the males have gone, as the eclipse plumage of the males of most species resembles the normal plumage of the female. More knowledgeable birders can still be confused, however, especially if confronted by a single individual bird, which can be a real challenge to identify. If all else fails, one can always watch and wait: eventually the bird will moult back into its full breeding plumage!

Shedding feathers

In all birds, the actual process of moult is more or less the same, though the order in which feathers are shed, and the time period in which this is done, vary enormously. What happens is this: the new feather gradually emerges from the skin, pushing out the old, worn one. Most birds shed their feathers one at a time, so that they still retain the powers of flight – vital for most in finding food. Others, such as ducks (see opposite), shed them all at once, and become flightless for a time.

Moult sometimes serves another purpose: enabling the males of some species (and the females of a very few) to adopt what is known as 'breeding plumage' – a rather brighter, showier set of feathers used in courtship displays.

OPPOSITE: Male Teal in eclipse – far less bright and colourful than his usual plumage!

Moulting stages

Most birds moult once a year; those that adopt a different breeding plumage usually moult twice; and one species, the Ptarmigan, moults three times a year – its plumage changing to help camouflage it against predators in its snowy environment. Young birds change their plumage very often in their very first year of life, going from a downy covering to their first full set of feathers (called juvenile plumage), and sometimes then into one other plumage (known in gulls as 'first winter') before finally attaining their adult feathers. In most species, this takes less than a year, but in some larger birds, notably large gulls and raptors, it can take up to four years to attain full adult breeding plumage.

BELOW: Both male and female Great Crested Grebes sport stunning head feathers during the breeding season.

ABOVE: Regular water bathing is essential to keep plumage clean and in good condition, as with this Oystercatcher.

OPPOSITE: Birds such as this Skylark often bathe in dust in order to remove grease from their feathers.

For birdwatchers, especially beginners, moult can be confusing, as birds undergoing a moulting process often look very different from their 'normal' appearance. However, for advanced birders, a good working knowledge of when and how different species moult can be a very helpful aid in identifying them; especially in complex groups such as warblers, large gulls and waders.

TIME FOR A BATH!

All birds bathe – though not all of them use water to do so. Some, from groups as varied as ostriches, bustards, bee-eaters, larks and sparrows, prefer to use dust as a way of cleaning oil and grime from their feathers.

Those that do use water have adopted a range of different methods in order to effectively wet their plumage without soaking themselves completely, which could leave them vulnerable to exposure in cold weather, or attack by predators at any time. Ornithologists have identified seven distinct ways in which landbirds bathe: standing on the edge of water and splashing themselves; crouching in the water up to their lower body; jumping in and out of the water; flying low over water and dipping their breast and belly feathers; plunging into water from a perch; deliberately exposing themselves to rain in order to soak themselves; and using the water on dense foliage or the dew on the ground in order to get wet.

Taking the plunge

Waterbirds such as ducks and gulls tend to bathe while swimming, contorting their head, neck and wings in order to get the water evenly distributed over the whole of their plumage. Others stand in shallow water and do the same. A few highly aerial species such as terns and frigatebirds plunge down into the sea, soak themselves briefly and then fly off before they get waterlogged.

Many birds bathe either once a day or several times, depending on the weather conditions and how dirty their plumage gets – those living in dusty environments such as deserts may need to do so more frequently. Birds dry themselves in a number of ways: some simply by shaking and ruffling their feathers in order to remove surplus droplets; others by sitting in the sun or simply in a warm place in the shade. Cormorants tend to hold their wings out to dry, giving them a very distinctive silhouette when perched. After bathing, many waterbirds will take the chance to re-waterproof their plumage using the special oil gland at the base of their tail, and preening themselves with their bill.

Dusting up

Dust-bathing, as already mentioned, is practised by a wide range of species, often those living in arid regions where water may be scarce. The methods used are quite similar to a bird bathing in water: crouching down and moving the body and wings in order to cover each feather tract effectively; then shaking to remove excess dust. It is thought that the grit not only helps remove excess oil on the feathers, but also helps get rid of parasites such as ticks and mites.

Without regular bathing, birds risk getting their plumage matted and oily, causing feather damage which may impair their ability to function – in particular, reducing their flight capability. Thus, just as with humans, bathing is an essential daily function – as important in the long term as feeding.

MIGRATION

Since early man first noticed flocks of birds passing overhead, migration has fascinated human beings. Yet despite our long study and observation of migrating birds, we are still far from knowing everything about their remarkable journeys and how they are achieved. Perhaps we should simply sit back and allow ourselves to marvel at these extraordinary feats of endurance, navigation and travel.

This chapter covers the various aspects of migratory behaviour, starting with the basic questions about how and why birds migrate, and how they find their way. It continues with a discussion of the various strategies used by different species on migration, and unusual migratory behaviour such as irruptions and atypical journey patterns. It also covers more local movements, the influence of the weather, and finally, the fascinating phenomenon known as vagrancy.

RIGHT: Snow Geese are strong fliers and undertake an epic twice-yearly journey between their Arctic breeding grounds and their winter quarters in the southern United States.

What is migration?

Migration is one of the greatest mysteries of the natural world and certainly the thing that most baffles, excites and intrigues human observers of birds. These epic seasonal journeys, from one part of the globe to another, are the stuff of dreams, giving rise to all sorts of questions. How do the birds find their way? Where do they come from and where do they go? And most intriguingly of all, why do they migrate in the first place?

WHAT, WHY AND WHERE?

Most migratory journeys, although by no means all of them, involve a twice-yearly journey: once travelling to the area where a bird breeds (usually taking place in spring), and once travelling back to their wintering grounds (usually in autumn).

However, there are exceptions to this rule, and not all migrations are as long-distance or, indeed, as regular as this phenomenon suggests. Some species migrate very short distances (perhaps simply moving up and down a mountain-side); others migrate just once a year; yet others travel in the opposite direction from what one would expect. But all undertake some form of journey, with the primary purpose of finding different places for breeding, as well as outside the breeding season.

Global travellers

Most bird migration takes place between the northern hemisphere (where the birds spend spring and summer courting, nesting and raising young) and the southern hemisphere (where they seek out places with plenty of food to keep them alive at a time when they might starve if they stayed put). Looking at the globe, it becomes apparent why most migration patterns fit this general scheme; it is because there is so much more land available in Europe, Asia and North America in which the birds can breed, compared to the relatively smaller area available south of the equator.

Thus, every spring, northern regions of the world see the return of millions – in fact, hundreds of millions – of species such as warblers, thrushes, flycatchers, chats, orioles, swallows and martins. And those are just the songbirds: raptors, storks, waders, swifts, cuckoos, bee-eaters and countless other non-passerine species also undertake this epic north-south migration.

LEFT: The migration of cranes has been known about for many thousands of years and was even mentioned in the Old Testament.

Between continents

From Europe and northern Asia, birds head south, either crossing the Sahara Desert or funnelling around its sides, in order to reap the benefits of plentiful food in sub-Saharan Africa. Other birds, from the northern regions of Siberia, head into India or travel south-eastwards to South-East Asia; some even go as far as Australia.

In North America, birds breeding in Canada, Alaska and the northern states of the continental United States mostly winter either in Central America, the Caribbean islands, or South America. Seabirds and waders from all around the Arctic tend to follow coastal or pelagic routes, some (such as the Arctic Tern) even reaching Antarctica.

Finally, in Europe and western Asia, many species of wildfowl do not head south, but instead travel in a westerly or south-westerly direction in autumn. Their destination is the wetlands of north-west Europe, especially Britain and Ireland, which are the winter quarters for millions of ducks, geese and swans due to the mild maritime climate of these islands.

Driven to survive

All these birds undertake such journeys for one simple reason: their chances of survival are greater than if they stayed within their breeding grounds. But why do they not simply remain on their wintering grounds all year round? The answer is do with the availability of food. If the birds were to stay, they would have to compete with the resident species for a finite food supply; but by heading north, they can exploit not only an abundant source of food, but also the long hours of daylight in which to collect it. It is fortunate for us that they do, because a world without the arrival of these birds each spring would be a far poorer one.

NAVIGATION AND ORIENTATION

Human beings have long marvelled at the apparently magical ability of birds to find their way across vast tracts of land and sea, in order to successfully complete their migratory journeys. Incredibly, even these days, when humanity has uncovered the answers to so many of life's mysteries, we cannot be absolutely certain how they do so with such uncanny accuracy.

What we do know, however, is that each migrating bird will adopt a combination of different strategies and tools, rather than relying on a single approach. We also know that birds are sensitive to the earth's magnetic field, and can use this to orientate themselves in a particular direction. Once underway, night migrants can also use the stars and moon to help find their way, while day-flying migrants can use the direction and aspect of the sun. On cloudy days they can also see polarized light, which helps them find their way; though this is probably less reliable than the solar compass.

Getting back home

Birds can also follow landmarks, which explains why geographical features such as coastlines, rivers and mountain ranges tend to be used as flight lines by large numbers of migrating birds. As they get nearer to their final destination, it is also thought that they can recognize the topography of the local area, enabling some species – including the Barn Swallow of Europe, Asia and North America – to return to the very place in which they were born.

OPPOSITE: Large flocks of migrating terns, often containing several species, are a regular sight along the coasts of all the world's continents.

Some birds travel in family groups or in large flocks, in which we know that the adult, experienced birds guide the youngsters, helping to reduce the mortality rate of birds undertaking their first migratory journey, and preventing them from getting lost. Yet we also know that in many species, including Arctic-breeding waders, the adults head south a few weeks earlier than the juveniles, leaving the youngsters to undertake journeys of thousands of miles with no guidance from their parents. Furthermore, the offspring of cuckoos never meet their parents at all, and so they always migrate on their own!

TIME OF YEAR

Different species have very different food, nesting and wintering requirements, which means that although outward migration takes place broadly in autumn, and return migration mainly in spring, these seasons are so extended that some birds, somewhere, are migrating at virtually any given time of year.

First arrivals

In North America and Europe, the first returning migrants tend to appear in February: sometimes early Sand Martins or Barn Swallows; or the odd Northern Wheatear. March sees the season really get underway: Chiffchaffs and Wheatears arrive in force, together with the first Blackcaps – once known as the 'March Nightingale' because its fluty song would confuse unwary listeners into believing that this later songster had returned. Sandwich Terns also come back in March, as do many species of seabird that have

LEFT: Each autumn, swallows and martins gather in flocks on telegraph wires to prepare for their long journey south.

wintered offshore or on the open sea.

April sees another rush: on this side of the Atlantic we get the main arrivals of swallows and martins, together with warblers such as Whitethroat and Sedge, Reed and Willow Warblers. Late April sees the first Swifts and Pied Flycatchers; while early May usually brings a major Swift arrival. Waders such as the Whimbrel are also April migrants, as is the Spotted Redshank. A few species come even later: the Spotted Flycatcher and the Turtle Dove tend to arrive in mid-May, while the tiny British population of Marsh Warblers may not return until the end of the month.

On the other side of the Atlantic Ocean the same pattern is followed, with waders and the first warblers passing along the east coast in April, and an even bigger rush in the first fortnight of May.

Return journey

By June, if one sees a migrant wader such as a Greenshank or Green Sandpiper, it is sometimes hard to tell if it is a latecomer heading north to breed or the first returning adult of the return migration. By July, any migrant is almost certainly returning south, heralding the start of autumn in the birds' – and birders' – calendar! Swifts and Cuckoos disappear by the end of July, with only a few stragglers hanging on beyond that date. Again, North America mirrors the European experience, with early migrants such as the Yellow Warbler, Louisiana Waterthrush and Willet heading south in August; while others delay their departure until September, October or even November.

In Europe, September is the main month for songbird and wader migration, with millions of birds passing through en route from Scandinavia, Arctic Russia and other northern locations to winter in Africa and Asia. Adult waders have often passed through earlier, leaving the juveniles to find their way on their own – their fresh, bright plumage a giveaway of their age. Large falls of songbird migrants such as warblers, flycatchers and chats sometimes occur on coastal sites and islands; especially where bird observatories have been placed to monitor the twice-yearly travels of birds.

By October, much of the migration is over, clearing the way for winter visitors to Britain such as geese and thrushes to arrive. In North America, thrushes are also late migrants, heading along the east coast from October through to November. Only in December and January does the primary southward migration reach a standstill.

Migration strategies

The notion that a tiny bird, weighing no more than a few ounces, might choose to migrate thousands of miles from the safety of its birthplace to an unknown destination in another continent is incredible enough; but when you realize that the bird may undertake virtually all its journey under cover of darkness, it seems even more incredible. Yet the majority of migrants, and certainly most migratory songbirds, do just that.

DAY OR NIGHT?

Night migration is particularly favoured by small birds such as warblers, flycatchers and chats, and for several very good reasons. First, small birds like these are very vulnerable to attack by predators; indeed, some raptors, such as Eleonora's Falcon, actually time their breeding in order to have their young in the nest during the peak of the southward autumn migration, when most songbirds are passing through the Mediterranean region. Most predators are diurnal, so by flying between dusk and dawn migrating birds at least reduce their chances of being attacked and killed.

Playing it cool

The other major reason is a physiological one. At night, the air is cooler than during the day, which means that not only can a bird fly more rapidly, but it also loses less water and uses less energy. Having flown

OPPOSITE: Tree Swallows are diurnal migrants, often seen hawking for insects over water.

BELOW: Yellow Warbler is one of the commonest summer visitors to North America, and is a nocturnal migrant.

How high?

Until relatively recently, it was hard to say how high migrants fly. However, in the past fifty years or so new ways of monitoring flocks of migrant birds have been discovered, including tracking by radar as well as visible observation. These methods have enabled us to create a clearer picture of what is happening.

The majority of migrants – especially songbirds – travel by night and tend to fly higher, and often faster, than day-flying migrants such as swallows and birds of prey. Most travel about 3,000–5,000 feet (1,000–1,500m) above the ground, though they can easily fly at heights of up to 10,000 feet (3,000m) at times. Waders, geese and swans can travel even higher than this, at heights of up to 20,000 feet (6,000m). This enables them to take advantage of rapid air currents in order to save energy, which is particularly important for heavy birds such as swans.

Other birds take a very different strategy: raptors will rise some two hundred metres or more on thermal air currents, then glide or soar as far as they can before finding another thermal; thus they tend to stay less than 1,000 feet (300m) above the ground. Swallows and martins fly very low indeed, skimming just above ground level and picking up insect food as they go. Seabirds such as albatrosses, petrels and shearwaters deliberately fly only a metre or so above the sea, taking advantage of the uplift caused by the passage of wind across the waves.

during the dark, the bird will then stop and rest at a convenient spot – often a desert oasis or an agricultural area with plenty of cover. Later on it will feed, before restarting its journey at dusk.

Many species, however, cannot migrate by night, either because they usually feed on the wing (insect-eaters such as swallows, martins and swifts), or because they are too heavy to be able to migrate without the aid of thermal air currents in order to gain height. Birds such as raptors (eagles, hawks and buzzards), cranes, storks and pelicans all migrate by day for this very reason.

OPPOSITE: Albatrosses are great global wanderers, often travelling thousands of miles across the southern oceans.

BELOW: Large birds such as cranes often fly in a 'V'-formation to reduce wind resistance while on migration.

Geese, swans and cranes

It is well known that many larger species of bird migrate in flocks, often flying in a 'V'-formation. The reason they do so is that this is the best way to save energy and to achieve the most efficient use of resources. By flying in a 'V', they are able to reduce wind resistance for the birds following behind the leader. Every now and then, the leader will change places with one of the following birds, so that a single bird never becomes exhausted.

This method is especially important for birds that have a high weight-to-wing area ratio, such as geese, swans and cranes, which would otherwise use too much energy flapping their wings.

Unusual migrations

Not all migrating birds head in the direction that one would expect, and for many reasons. Some follow ancestral paths which appear to have remained unchanged for tens of thousands of years, despite the fact that there would seem to be alternative routes that might appear more logical.

ODD DIRECTIONS

Take, for example, the Lesser Whitethroat and the Red-backed Shrike. Both species breed across much of Europe and western Asia, yet both species winter mainly in East Africa. So even though we might expect that populations breeding in western Europe would skirt around the western side of the Sahara Desert, in fact they all head eastwards in autumn, travelling around the eastern end of the Mediterranean and into Africa by this route.

The Aquatic Warbler, a rare species breeding mainly in the damp, sedge-filled meadows of eastern Europe and western Russia, does virtually the opposite. In autumn, instead of heading south, as might be expected, Aquatic Warblers travel westwards along the coasts of the Low Countries and France, before turning southwards and heading into western Africa. As a result, a few Aquatic Warblers are found in reedbeds in Southern England each autumn, though because of their skulking habits they are usually trapped by ringers rather than spotted in the field.

In fact, this species is so difficult to see that we do not know where exactly its winter quarters actually are, but we have reason to assume that the birds spend the non-breeding season somewhere south of the Sahara.

RIGHT: Curlew Sandpipers undertake a twice-yearly journey between their Arctic breeding grounds in Siberia and winter quarters in Africa.

Passing through

Another species that does not breed in Europe is a much more regular passage migrant through Britain, especially in autumn. The Curlew Sandpiper breeds no nearer than the Yenisei Delta in Siberia, about 90 degrees east (or one quarter of the way around the globe). Yet because the species winters throughout sub-Saharan Africa, many birds head south-west in autumn, passing through coastal sites in Britain. In contrast, the Broad-billed Sandpiper, which breeds in Norway, is a very rare vagrant here, with only a handful of records each year.

At least these birds are heading in a broadly southerly direction, even if they do take a diversion or two along the way. But one species of passerine actually migrates north to spend the winter in Britain. The Water Pipit breeds in the mountainous areas of central and southern Europe, and while some birds (notably those in the Alps) simply head to lower altitudes in autumn, others fly several hundred miles northwards to spend the winter on the

watercress beds and freshwater marshes of southern England and East Anglia.

LONG-HAUL TRAVELLERS

When it comes to long-haul travellers, birds have the rest of the animal kingdom beaten. Indeed, their journeys are even longer, in some cases, than our own long-distance flights across the globe – and they have to do it entirely under their own steam!

Champion travellers include the Arctic Tern, which may, during the course of its twenty-year lifetime, be the world's most travelled living creature – flying as much as half a million miles.

Arctic Terns, as their name suggests, breed mainly in the higher reaches of the northern hemisphere, including Iceland, Scandinavia and the British Isles. They do so because of the abundance of fish in the seas around this area, and also because of the long hours of daylight during the summer months, which means they can raise a family. Having done so, in autumn the terns head south, along the western coasts of Europe and Africa, travelling at least 11,000 miles to their winter-quarters around the Antarctic pack-ice. Here, they enjoy equally long hours of daylight and an even more abundant source of food.

Ocean wanderers

Although the Arctic Tern is the undisputed champion long-distance migrant, it is not the only bird that undertakes incredible journeys. Other ocean-going seabirds such as albatrosses and shearwaters regularly travel thousands of miles from their breeding grounds on remote islands, in search of food for themselves and their young.

Birds crossing the vast Pacific Ocean, which covers almost half the earth's surface with only a few scattered islands amidst the stretches of open sea, face huge challenges. Many Arctic breeding waders spend the winter on archipelagos such as Fiji or New Zealand, and to reach their destination have to cross large expanses of open ocean in a single hop. Bar-tailed Godwits and Pacific Golden Plovers are two species that do so, the godwits travelling almost 7,000 miles, sometimes without resting.

Among songbirds, two species share the prize for the longest journey. Barn Swallows travel from Scandinavia and northern Britain to winter in southern Africa, a distance of about 6,000 miles. The Greenland subspecies of the Northern Wheatear must travel from its northerly breeding grounds, via western Europe and across the Sahara, an equally long and arduous journey.

LEAPFROGS AND LOOPS

Not all migratory journeys are the same in spring as in autumn. In North America, millions of songbirds set off each autumn from their breeding grounds in Canada. They take advantage of tail winds provided by a cold front in order to take the shortest route south, which entails flying for several thousand miles over the open ocean of the west Atlantic. This has several advantages: not only does it reduce the time taken to travel to their wintering grounds, but it also reduces the chances of attack by land-based predators, such as hawks. The downside is that if the birds encounter stormy weather conditions, such as the tail-end of a hurricane, then many birds will fall into the sea and perish, unable to return to land in the strong winds.

In spring, birds such as wood-warblers, thrushes and vireos must follow a land-based route, taking them from the Caribbean or Central America up the eastern seaboard of the United States, before heading inland to their breeding grounds. They follow this route because, unlike in autumn, when the weather can help them, in spring the winds tend to be against them, making an ocean crossing impossible.

OPPOSITE: The Arctic Tern holds the record for the longest migration distance: a round trip of at least 11,000 miles.

BELOW: Bar-tailed Godwits are great global travellers, making huge flights across the Pacific Ocean.

Different routes

In Europe, several species follow a different route in spring than in autumn. Seabirds such as Pomarine Skuas often fly up the English Channel on their way to Scandinavia; whereas in autumn they take a different, more continental route; a process known as 'loop' migration. In some species, adults and young birds take a different route: juvenile Dunlins migrate from Russia to eastern Britain via Norway, whereas their parents take a more southerly route via the Baltic, so that they can stop to moult from their breeding to their non-breeding plumage at the Wadden Sea, in the Netherlands.

Other species of wader follow a pattern known as 'leapfrog' migration, whereby the most northerly breeding populations winter the farthest south. So Ringed Plovers breeding in the Arctic will winter off the coast of Africa, whereas those breeding in Scotland may only travel as far as southern Britain.

IRRUPTIONS

Not all birds make the same regular twice-yearly migratory journeys; some, known as 'irruptive' species, make less regular movements, depending on the availability of their preferred food. This tends to apply to species that breed in northerly latitudes of Europe or North America, whose food supply is not always guaranteed.

One of the best-known irruptive birds is the Bohemian Waxwing, which occurs in both Europe and North America. Waxwings feed on insects in summer, but in autumn and winter they need a good supply of berries to see them through the season. The berry crop is cyclical, with good years and bad, so from time to time food shortages force large numbers of Waxwings to head southwards in search of food.

On the way

In Britain, the first sign that a Waxwing invasion is on its way usually comes in early autumn, when the first flocks appear on the east coast. They are often found in what might seem to be unusual places, including gardens, parks and especially supermarket parking lots, which often have a profusion of suitable berry-bearing bushes.

Once they have arrived, flocks gorge themselves on berries before heading on, so that by late winter they may have spread to different parts of the country. Each year, birders hope that some will stay on and breed; but some ancestral impulse drives them back to their usual homes of

LEFT: Some populations of Ringed Plover travel further than others, 'leap-frogging' their neighbours in the process.

OPPOSITE: Snowy Owls occasionally roam far to the south of their normal range, searching for food.

Shearwaters

Two species of seabird do the opposite of the vast majority of global migrants, and breed in the Southern Hemisphere but spend the southern winter (our summer) in the Northern Hemisphere. Great Shearwaters mainly breed on three remote oceanic islands in the South Atlantic: Inaccessible Island and Nightingale Island (both part of the Tristan da Cunha group), and Gough Island; meanwhile, their relative the Sooty Shearwater breeds in vast colonies off the coasts of southern South America and the South Atlantic islands.

After breeding, these birds head northwards on a vast loop: over the Equator, along the eastern seaboard of North America, then across the Atlantic. In July, August and September they can be seen in varying numbers off the British and Irish coasts, though Great in particular is often so far offshore you need a boat to find them. Great Shearwaters occasionally occur in large numbers too: thousands have been seen off Cornwall and southern Ireland on rare occasions, suggesting that there must be large flocks out in the open ocean.

Sometimes large numbers of Sooty Shearwaters occur in the North Sea, where they can be seen from east coast headlands. But by October, they have left our waters and begun the long journey south, arriving back to the Southern Hemisphere in time to take advantage of the austral summer to breed.

Scandinavia and northern Russia.

Another group of birds that also undergoes periodic but irregular irruptions are the various species of crossbill, also found across northern Europe and parts of North America. Crossbills tend to irrupt in late summer, in search of new areas of pine and spruce cones, where they can feed. Unlike Waxwings, once they have found a new home, they tend to stay to breed there; though being very early nesters, their breeding season begins as early as January.

Rise and fall

Other irruptive species include some of the 'boreal' owls, which live in the tundra and forests of the north. Snowy Owl populations rise and fall in line with those of their preferred food, lemmings; when there are shortages, the owls head south, turning up in northern Scotland and in New England.

Local movements

Many birds which do not normally migrate nevertheless undertake localized movements in autumn and winter, governed by changes in the weather. Some move out from high ground as winter arrives, looking for warmer areas lower down; cold snaps may result in sudden departures and wide dispersions.

HARD WEATHER MOVEMENTS

Recent climate change means that hard winter weather has become something of a rarity in western Europe; during severe snow and icy conditions, birds respond by heading southwards and westwards in order to avoid the problem.

Some species are more prone to hard weather movements than others. Geese, ducks and swans have often been linked to weather forecasting, because flocks often turn up in autumn just before the weather changes to colder conditions. Other reliable indicators of cold weather to come are Lapwings, which tend to respond to changes in pressure between twenty-four and forty-eight hours ahead.

Unusual crossing

Lapwings tend to head from eastern Britain into the south-west, or across the Irish Sea to Ireland; but if caught up in unusual weather systems they may even cross the Atlantic Ocean. In 1927 hundreds were blown off course by an easterly gale and ended up in Newfoundland, in eastern Canada; though occasional vagrants have crossed the ocean since, they have never done so again in such large numbers.

Other hard weather migrants include two species of winter thrush, Redwing and Fieldfare, which breed in Scandinavia and northern Europe but winter in large numbers in Britain. Redwings in particular are very vulnerable to cold weather, and when bitterly cold spells occur they will head across the English Channel to the continent, in search of milder conditions.

Ducking out

Waterfowl – especially ducks and geese – may respond to hard weather in the Netherlands by heading west to the reservoirs and gravel-pits of southern Britain; though again, as a result of the trend towards milder winters, this is becoming a less regular occurrence than before. Species involved include Smew, a species

LEFT: Lapwings are especially sensitive to harsh winter weather and will try to move ahead of oncoming snow.

of duck that winters in large flocks on the Dutch polders, but is only found in small numbers in Britain, mainly in the south-east.

Winter gales can also wreak havoc on flocks of seabirds, many of which spend the winter out in the open sea. Such incidents are known as 'wrecks', and can involve large numbers of birds such as Guillemots and Razorbills being driven onshore, where they perish. In a few cases, gales may even drive these birds far inland; some manage to reorient themselves and find their way back to the open sea, but many have to be taken into care and then returned to the coast, where they are released.

ABOVE: The Ptarmigan is an altitudinal migrant, moving seasonally up and down the mountain.

OPPOSITE: European Bee-eaters are seen increasingly in Britain in spring, having 'overshot' their intended destination.

Altitudinal migration

Not all movements cover vast areas of the globe; yet for the birds concerned, even short journeys may be vital to their survival. Several montane species – those which habitually breed at high altitudes – undertake what are known as vertical or altitudinal migrations, in which they move to higher levels in spring and summer to breed, and to lower levels in winter. Such movements are, like all migrations, basically governed by the availability of food.

Thus in Britain and Europe, birds such as the Ptarmigan, which only breed on high mountain tops, move several thousand feet lower in autumn. Other species which breed at both high and low altitudes, such as Meadow Pipit and Skylark, also move away from hills and mountains in winter, so that areas of the Scottish Highlands can be virtually devoid of birds during this season. Many head towards coastal areas, which tend to have a milder climate and, consequently, more food.

In North America, several species undertake very localized movements between summer and winter, with Mountain Quail carrying out what must be the shortest migratory journey of all, travelling a few thousand feet down the mountain on foot!

Birds and weather

Birds are at the mercy of the weather. Bad weather can disrupt the long migrational journeys that some birds undertake; good weather will help birds to travel, but can mean that some birds fly too far in spring. Severe and unusual weather conditions can lead to deaths on a large scale, but might also bring vagrants to our shores.

SPRING OVERSHOOTS

Every spring, millions of migrating birds return to western Europe from their winter quarters in Africa south of the Sahara. Some, like the Knot and Bar-tailed Godwit, travel as far as the Arctic, taking advantage of long hours of daylight and abundant food supplies in order to nest and raise their young. Many, including songbirds such as Pied Flycatchers, Willow Warblers and Whitethroats, will breed in more temperate latitudes. But other species, including several kinds of heron and egret, and exotic species such as European Bee-eater and Black-winged Stilt, stop off in the Mediterranean region, with most of the population breeding there.

A flight too far

But in some springs, birders in southern Britain are treated to small numbers of these essentially southern European species appearing unexpectedly on our shores. This phenomenon – known as overshooting – occurs during settled periods of weather, when a large area of high pressure sits for a while over western Europe. High pressure systems mean clear skies and light winds, usually blowing from a southerly direction. Under such conditions, returning migrants heading north from Africa sometimes keep on flying, not making landfall until they have crossed the English Channel.

Drift Migration

Another spring and autumn migration phenomenon is more complex and harder to predict. Drift migration is a term first coined in the 1950s, when birdwatchers and observatory wardens on the east coast of Britain noticed that falls of migrants often coincided with light easterly or south-easterly winds, and were not necessarily linked with bad weather (the cause of the largest falls of migrating birds, which get disoriented in heavy rain and high winds).

This theory suggests that in autumn, birds leave Scandinavia in good weather, with light following winds. However, at some point as they are crossing the North Sea they encounter easterly winds, which push them westwards. As they approach the coast, they may come upon clouds or even rain, which disorient them still further and force them to seek landfall as quickly as possible. Hence the arrival of large numbers of migrants on the east coast of Scotland and England, especially at migration watchpoints such as Fair Isle, the Isle of May and Spurn Head.

Drift migration is less common in spring, but can occur when high pressure over eastern Europe pushes migrating birds heading for Scandinavia westwards. These may allow themselves to be drifted in this direction until they make landfall, then feed and replenish their energy resources before heading back across the North Sea. Species involved include the Red-backed Shrike, Icterine Warbler and Bluethroat.

Species involved in such overshoots include several kinds of small heron: Night Heron, Squacco Heron and Cattle Egret being the most frequent, along with the occasional Little Bittern, a skulking species that is often hard to see. Hoopoes and European Bee-eaters are other frequent overshooters, the latter sometimes arriving in small flocks, as it is a sociable species which likes to migrate in groups. Other Mediterranean species include landbirds such as the Great Spotted Cuckoo, Alpine Swift and Woodchat Shrike.

Overshoots can occur as early as March (the time when some species arrive back to southern Europe to breed), or as late as May (when non-breeding birds, which are perhaps more likely to wander, will return north).

OPPOSITE: The Red-footed Falcon is a rare vagrant to Britain, usually when there are easterly winds in spring.

ABOVE: Hoopoes are a rare and irregular British breeding bird, but may colonize as a result of global warming.

Opportunist breeders

Some of these birds occasionally stay to breed in Britain, with Hoopoes being the most frequent – though still an irregular – breeder. European Bee-eaters have attempted to nest four times, though they have only been successful twice, in 1955 and 2002. Black-winged Stilts have a similarly patchy record as British breeders, with pairs nesting in the 1940s and 1980s, and attempting to breed at other times.

It is possible – even quite likely – that eventually a spring overshoot will allow one of the smaller herons to colonize Britain, the most likely contender being one of the world's most successful birds, the Cattle Egret, which often migrates in small flocks.

Spring overshoots are also regular in North America, where large raptors such as Swallow-tailed and Mississippi Kites regularly appear much farther north than their southern breeding grounds, occasionally reaching New Jersey.

DEATH IN WINTER

Hard winters may appear to be a thing of the past in Britain due to the effects of global warming, yet it was not so long ago that severe winters brought death to millions of birds. Winters such as those of 1916–17, 1940–41, 1946–47 and the 'Big

ABOVE: Kingfishers are very vulnerable to harsh winter weather, as icy conditions make it impossible for them to feed.

OPPOSITE: Benefiting from a recent run of mild winters in southern Britain, the Dartford Warbler has extended its range.

Freeze' of 1962–63 blanketed the country with snow and ice for several months, and brought massive increases in winter mortality rates.

Birds die in winter for one basic reason: they cannot get enough food. Thus they lose energy and weight until they eventually become too weak to feed or fly, and then they die. Cold itself is not a major killer: birds can survive very low temperatures, just as long as there is food to eat. So the recent trend towards feeding birds in our gardens has undoubtedly contributed to the survival of many birds – especially juveniles undergoing their very first experience of the hardships of winter.

Deep freeze

Water, too, is a priority for birds in winter, especially when temperatures drop below freezing and natural sources of water such as puddles and ponds freeze solid. Again, by putting out a regular supply of water householders are contributing to bird survival.

Short cold spells, lasting a day or two at the most, are relatively unproblematic for birds. It is the freeze-ups that go on for a week or more that have a major effect. Some birds choose to flee the coming weather, heading southwards and westwards to winter on the European mainland or in the milder south-west of Britain. Others do not seem to mind the cold: scavengers such as kites, crows and buzzards, and predators such as the Sparrowhawk or Tawny Owl, do well in cold, frosty conditions.

Small casualties

The major casualties of prolonged cold spells in winter are small birds, especially those that feed on insects. Most insect-eating species migrate south to warmer climes for the winter, as survival in northern regions is difficult. But some – especially those that glean insects from the bark of trees, such as the Goldcrest, Long-tailed Tit and Treecreeper – choose to stay. In normal winters they do quite well, so long as they know where to look for food. But hard weather, especially rain followed by a freeze, which covers twigs and branches with a thin but impenetrable layer of frost or ice, makes it impossible for these birds to find food. With no alternative food sources, many will die.

Waterbirds are also very vulnerable to cold weather. Birds that wade to feed, such as Grey Herons, need to seek out alternative sources of food, often visiting garden ponds in order to plunder the goldfish. Kingfishers are also highly vulnerable, as they can only feed by plunging into water. When rivers and streams freeze over, as happens in the most severe winters, they have to migrate or they will die.

Bouncing back

Hard winters do have a short-term effect on the population of vulnerable species, with more than 90 per cent dying in very bad years. However, most appear to be able to bounce back, as the tale of one common and one rare species shows. In the terrible winter of 1962–63, many millions of birds died, with Wrens being particularly hard hit. At the time, the Wren was in the top ten of Britain's commonest birds, but far from being in the number one spot. Yet within a few years the Wren population had bounced back, and by the turn of the millennium it was Britain's commonest bird, easily outstripping its closest rivals.

The reason for the Wren's comeback is that since 1963, apart from a couple of colder than average winters in the 1980s, Britain has experienced an unprecedented run of very mild winters. As a result, far fewer young Wrens die of starvation each year, leading to a population boom, with the combined numbers of breeding pairs for Britain and Ireland soon to overtake the ten-million mark.

Another, much rarer species, the Dartford Warbler, also suffered badly during the 1962–63 winter, with numbers falling as low as a dozen breeding pairs, all in Dorset and the New Forest. Despite its name, the Dartford Warbler is a mainly western European species, with strongholds in France and Spain, and on the northern edge of its range in Britain. However, thanks to the run of mild winters, this resident warbler is no longer rare or endangered as a British breeding species. It now thrives on the heaths of southern England, from Cornwall to Sussex, and some birds have colonized the coastal heaths of Suffolk. Who knows, given time and more climatic amelioration, they may even recolonize the Dartford area of Kent, where the first specimens were shot and collected over two hundred years ago.

Vagrancy

Nothing excites birders the world over more than a bird found well outside its normal range: a bird that should not be where it is at all, but has arrived at its destination either by becoming lost, having been deflected off course by bad weather, or by a form of wanderlust which prompts birds to explore new areas, which they may eventually colonize.

VAGRANTS, WAIFS AND STRAYS

Vagrants can turn up anytime, anywhere, and from any place, but they tend to be seen mostly during the two migration seasons of spring and autumn. Coastal areas are often vagrant hotspots, especially if they are on migration flyways or isolated from other areas of land: for this reason remote islands tend to concentrate vagrants. Rare birds also typically appear where there are birders already looking for them, which means that good places for watching vagrants tend to get better, while a rare bird may never be noticed in other areas that are less well-watched.

From all around

Britain and Ireland are uniquely placed to receive vagrant birds from all over the globe, being on the edge of a large land-

OPPOSITE: Allen's Gallinule from Africa has been seen several times in Europe.

ABOVE: Ross's Gull is a rare – and always welcome – visitor from the Arctic regions.

mass, with migration routes crossing the area from all directions. Thus, in spring we get rare visitors from southern and eastern Europe, such as bee-eaters and Rollers; in autumn we see visitors from the north and east, even as far as Siberia; and from the west, visitors arrive when transatlantic storms and gales bring vagrants from North America.

But rare birds also visit us from other directions: Ivory and Ross's Gulls come from the Arctic, Allen's Gallinule from sub-Saharan Africa; from south-east Europe comes Rüppell's Warbler; the Pacific brings the Ancient Murrelet and Aleutian Tern; and the South Polar Skua even arrives from Antarctica. Indeed, rare birds arrive in Britain from all over the place, making it the rarity capital of the world.

Head for the coast

Birders keen to seek out rare birds in Britain and Ireland tend to head to coastal headlands and islands, many of which are or have been bird observatories. Fair Isle, lying between Orkney and Shetland, has seen the highest number and variety of rare vagrants, especially in September and October; while the island of Lundy off the Devon coast is also a major hotspot. Dungeness in Kent, Blakeney Point in north Norfolk, Spurn and Flamborough Heads in Yorkshire, and Cape Clear Island off County Cork in southern Ireland have also had their fair share of rare sightings over the years; as have the Isles of Scilly, which play host each autumn to an annual invasion of keen twitchers, each eager to add more rare species to their life list.

Short stay

Most vagrant birds only stay put for a day or two, perhaps a week at most, though occasionally they may be present for months or even years, such as the Glossy Ibis which remained in Kent for well over a decade. Others, especially transatlantic vagrants such as Ring-billed Gulls, return to the same site each winter, presumably migrating north and south with their close relatives, and spending the intervening breeding season with them.

Vagrancy is closely linked both to migration patterns and to the weather. Thus, for example, it is thought that many vagrants from Siberia to Britain have a defective compass, which sends them west instead of east in autumn. Conversely, vagrants from North America are often blown eastwards by gales, and end up making landfall in south-west Britain or Ireland after a long and arduous flight.

The most sought-after species are first-ever sightings for Britain and Ireland, many of which originate from as far afield as North America and Siberia.

But inland sites also get their share of rare vagrants. Indeed, a while ago the inland county of Nottinghamshire played host to two 'firsts' from North America: Cedar Waxwing and Redhead, a type of duck closely resembling a Pochard. Because inland sites are often watched regularly by local birders, such sightings are becoming increasingly frequent.

ACROSS THE ATLANTIC

There are well over fifty species on the official British List that do not even breed in the Old World continents of Europe or Asia, but are confined to the New World. These birds are true vagrants which, due to a combination of their migratory habits, the weather and a stroke of luck (good or bad, depending on one's perspective!), have crossed the Atlantic Ocean to the delight of birders and twitchers.

These include a wide variety of families, most commonly the American wood-

OPPOSITE: The Cedar Waxwing from North America has been seen only a handful of times in Britain.

ABOVE: Wilson's Phalarope is one of the most frequent New World wader species to stray across the Atlantic to Europe.

warblers and waders, but also grebes, vireos, thrushes, sparrows, buntings, terns and gulls. Not all arrive here at the same time of year or for the same reasons, but in most cases unusual weather is involved.

Waders from the west

Waders are among the most long-distance migrants and are strongest flyers of all birds, so it is hardly surprising that more than a dozen North American species have been recorded on this side of the Atlantic, with several species, such as Pectoral and Buff-breasted Sandpipers, Lesser Yellow-legs and Wilson's Phalarope, turning up in fairly good numbers. These species migrate out over the Atlantic from their Arctic breeding grounds, heading for Central or South America. They tend to fly fairly high, and sometimes get caught up in high altitude winds, along the path of the jet stream (which also enables aircraft to make quicker time crossing from west to east than in the other direction).

Waders are hardy birds, and so many survive for some time after they arrive, sometimes being recorded the following spring. However, due to the prevailing westerly air stream and weather systems, it is highly unlikely that any make it back to their original homes. In the 1970s, one pair of American waders, Spotted Sandpipers, even nested in Scotland.

Unlikely to survive

Smaller species such as wood-warblers and thrushes are also regularly blown across the Atlantic, mainly during gales in

ABOVE: *Despite its dumpy appearance, the Pied-billed Grebe is a strong flyer, able to cross the North Atlantic Ocean.*

late September or October. These tiny waifs and strays are unlikely to survive for very long here; as soon as the temperature drops and their food supply disappears, they are doomed. Some highly unlikely species cross the Atlantic, such as the Pied-billed Grebe, a poor flyer which nevertheless has managed to do so several times. Other unlikely transatlantic travellers include Britain's only American Purple Gallinule, picked up exhausted on the Isles of Scilly in autumn 1958; a Red-breasted Nuthatch which frequented a coastal pine wood in Norfolk during the 1980s, and a Yellow-bellied Sapsucker, a kind of woodpecker which appeared on Scilly in 1975.

OPPOSITE: Pallas's Warbler regularly turns up in western Europe, thousands of miles from its intended destination.

Ship assistance

Back in October 1962, a passenger on a ship crossing the Atlantic Ocean witnessed an extraordinary sight: that of hundreds of small birds landing on the decks, masts and funnels of the vessel. Most disappeared soon afterwards, heading south in an attempt to reach landfall in the Caribbean. But a few stayed with the ship for the whole journey; and could be seen flying ashore as it approached the docks in Southampton a few days later.

This was the first major recording of a controversial phenomenon known by birders as 'ship assistance'. It is controversial, because the debate still rages as to whether or not you can 'count' as a genuine British sighting any bird which may have taken a rest on an ocean-going vessel on its way across the Atlantic. Not so long ago any record of an American landbird was rejected because of the possibility that it had hitched a ride; today, however, the rules have been changed so that as long as the bird has remained in a wild state, and has not been fed, assisted or taken into captivity by the crew of the ship, then it is deemed to have reached Britain under its own steam.

This has led to many birds sighted around the area of ports, such as Liverpool in Merseyside or Felixstowe in Suffolk, being accepted as wild despite some purists' reservations. But as some twitchers like to say, birdwatchers can count what they like – so long as it makes them happy!

Abmigration

Abmigration is defined as the phenomenon whereby a particular individual from one species accidentally joins members of another species, and instead of migrating to its usual wintering or breeding grounds, ends up following its new companions to their usual ones. This phenomenon is especially common in birds that migrate in large, sociable flocks, such as geese.

Thus Snow Geese have been known to latch on to flocks of the Greenland race of White-fronted Geese in Arctic Canada, and instead of heading south with their fellows to Texas they have instead travelled with the White-fronts to their wintering grounds in Europe. Most years, one or two vagrant Snow Geese appear on the Hebridean isle of Islay, feeding all winter before presumably returning home in the following spring.

In recent years, numbers of two rare Asian species, the Yellow-browed and Pallas's Warblers, have been sighted on the east and southern coasts of Britain in autumn. It is now thought that they may now be wintering somewhere in West Africa instead of in South-East Asia, having changed their behaviour as a result of their global wanderings. Again, this may have been due to individuals joining forces with other migrants and heading in a different direction than usual.

FEEDING

Like us, birds need to eat regularly in order to gain enough energy to survive. The way in which each bird feeds – and the variety of foods that they choose – is a fascinating area of bird study. It is also one of the easiest to observe for oneself: anyone can do so by simply putting food out for birds in a back garden.

This chapter begins with a general introduction to the food that birds eat, followed by a look at the different feeding methods they use and the ways in which these have shaped (literally, in many cases) their body forms and behaviour. A look at some unusual feeding strategies such as kleptoparasitism (also known as piracy) is followed by a look at the complex relationship between predators and prey. A section on specialized feeders and their feeding methods is followed by a brief look at the ways in which birds drink.

RIGHT: The consummate hunter, the Barn Owl is able to pinpoint its prey through its extraordinary powers of hearing.

What do birds eat?

Birds vary their diet considerably – from family to family and even from individual to individual. Some birds mainly eat one particular type of food, while others will eat almost anything. Many species have very specialized diets which dictate their feeding behaviour, and food availability may also influence how they feed.

FOOD FOR FREE

Some birds tend to eat a single kind of food: fresh meat for many raptors, fish for seabirds, flying insects for swallows and martins. Others vary their diet from season to season. For example, Blue Tits will feed on seeds and nuts during the winter (especially those provided by us), but during the breeding season they feed their young on caterpillars. Others change their diet over a longer period of time: for example, Robins have learned to come to feeders in gardens, and sometimes join the tits and finches on hanging feeders where they eat sunflower seeds or nuts, not part of their normal diet. Blackcaps wintering in Britain have also become less specialized, happily feeding on insects in

BELOW: Pelicans are formidable hunters, corralling fish together before grabbing them with their huge bills.

summer but varying their diet to include nuts and seeds in winter, when the insects are no longer available.

One of the richest and most productive sources of food on the planet is the open sea, yet paradoxically the ocean can also be the place where it is hardest to find food, given the concentration of edible resources in certain areas. Nonetheless, the ocean provides a wide range of food for birds, from tiny plankton to large fish, supporting a wealth of seabird species.

Wide tastes

Some species have become almost completely omnivorous: many species of gull, which used to survive in the wild mainly by eating fish offal scavenged from over the side of fishing boats, now depend almost entirely on refuse at landfill sites. Others have even been known to dive down on unsuspecting passers-by and take Cornish pasties, ice-creams and even sweets!

Many species retain their specialized feeding habits and diets. Most raptors (with a few obvious variations, such as the Osprey, which feeds on fish) eat meat, either killed by themselves or scavenged. Hummingbirds are ultra-specialized, feeding on nectar which they obtain from flowers using their specially adapted bills. Other specialist feeders include many types of wader (eating mainly crustaceans, worms or other aquatic creatures); parrots (fruit) and vultures in both the Old and New Worlds (rotting meat scavenged from the corpse of an animal).

Getting stuffed

Birds time their feeding very differently, too. Vultures will gorge themselves on as much meat as they can eat at one sitting, then rest for several days, digesting their meal. Only then will they need to feed again. In complete contrast, small birds may need to feed every few seconds, especially during the cold winter months, when daylight hours are short and they need enough energy to survive the long, dark night. Thus a flock of tits will move through a wood in winter at a frantic pace, picking at tiny insects on or under the bark of trees.

Feeding methods vary considerably as well. Foraging is the commonest method: simply moving through a suitable habitat picking up food items as they are found. This approach is used by birds where the food supply is abundant but comprises small items, so that the bird needs to feed more or less constantly. This method is used by such diverse groups as waders, ducks and warblers. Conversely, raptors and owls often kill their prey, and may spend a considerable time in the chase.

Feeding methods and strategies

Birds have developed a wide range of feeding strategies in order to catch their prey. These are reflected in all sorts of aspects of birds' physiology and anatomy, and of course in their behaviour. In some cases, the feeding opportunity that a bird exploits has been a key factor in the way a group of birds will diverge and evolve in order to fill separate ecological niches.

ODD-SHAPED BILLS

Some species – and, indeed, whole groups or families – have evolved unusually shaped bills, so that they can best exploit a particular food supply. Waders are one such group, with one of the most diverse range of bill shapes of any bird family.

Starting with the larger species, there are the curlews, whose bill curves downwards in order to probe deep into the mud most effectively; and the avocets, whose bill curves upwards in order to be most effective when swept from side to side through shallow water, filtering out tiny aquatic organisms. Snipe, too, have very long bills, designed to probe vertically down into mud, with tiny sensors on the tip to allow the bird to detect its prey.

Plovers, such as the Lapwing, tend to have a short, stubby bill, which they use as a general-purpose instrument to pick food off the surface of the mud. Other small waders such as the Dunlin or the stints have short (sometimes decurved) bills, which they again use either to probe or to pick items off the surface.

BELOW: The Avocet feeds by sweeping its upturned bill from side to side to grab minute aquatic insects.

OPPOSITE: Seabirds such as gulls often follow fishing trawlers to pick up cast-off food.

Seabirds

Seabirds have developed a variety of different methods and strategies in order to avoid going hungry. The simplest is to follow fishing boats, and this is mainly – though far from exclusively – practised by species and groups which live along the coast, rather than venturing out into the open sea. Gulls are the prime exploiters of this by-product of our own need for food, and learn to time their visits to the boats at exactly the same time as the fish are being gutted and the waste products are being thrown over the side. Other birds often seen around inshore fishing boats include Gannets and skuas.

When fishing boats venture farther afield – perhaps 20–30 miles offshore – they begin to attract what birders call 'pelagic' species, such as shearwaters and petrels, and in the southern hemisphere, albatrosses. These have an extraordinary ability to find food using their sense of smell, though vision also plays a part, as they scan the horizon for concentrations of birds which usually indicate a productive source of food. Ocean-going voyagers are often amazed at how these birds seem to appear from nowhere, follow a ship for a while, then vanish in search of a more productive source of food.

Scavengers

Not all birds of prey kill their own food. Vultures – both Old and New World species – are exclusively scavengers, taking advantage of the hard work done by others! They may be joined at the kill by other scavenging birds, especially Marabou Storks (Africa), kites (Europe, Africa, Asia and Australasia), caracaras (South and Central America) and crows (more or less everywhere!).

Despite their superficially similar appearance, the Old and New World Vultures are in fact totally unrelated, the former having a common ancestor with eagles, hawks and buzzards, the latter descending from an ancestor of the storks. This different heritage shows in the very different methods they use to find food: Old World Vultures such as the Griffon locate it mainly by sight, whereas New World species such as Turkey and Black Vultures do so using their well developed sense of smell, which Old World species lack.

The reason that these very different methods evolved is that the habitats where these birds search for food are very different. In Africa, they generally hunt across vast areas of savannah, where sight is more useful than smell; whereas in the Americas, they usually find food in dense jungle, where smell is more important.

Weird waders

Occasionally, one particular species will develop a really peculiar shaped bill; in the case of the waders, it is perhaps not surprising that several species have done so. The two most unusual are that of a New Zealand relative of the plovers called the Wrybill, and an east Asian relative of the stints, the Spoon-billed Sandpiper. The Wrybill's bill is the only one in the world of birds to curve from right to left; enabling this little wader to prise small creatures from beneath stones. The Spoon-billed Sandpiper's bill, as its name suggests, is rather like that of the spoonbills: a spatulate shape which helps it filter food. The Shoveler, a species of duck , also has a flattened, spatulate bill, which it uses for a similar purpose.

The bill of the various species of flamingo is arguably the most specialized of all. Flamingos have a peculiarly shaped bill: decurved, with a bulbous nose, designed to hold their large, fleshy tongue.

OPPOSITE: Vultures are the quintessential scavengers, feeding on the carcasses of large mammals.

ABOVE: The Black Heron is also known as the 'umbrella bird' because of its unusual feeding style.

When feeding, flamingos hold their bills upside-down and dip them into the water, then partially open the bill and waggle their tongue, filtering tiny aquatic organisms into the bill. Amazingly, the only other creatures that use a similar feeding method are the baleen whales!

FEEDING STRATEGIES

Apart from the foraging behaviour previously described, birds can pursue a variety of active and passive ways to feed. Among the passive approaches is the 'wait-and-see' strategy, which is practised by a range of species, especially those that feed in rivers or other waterways. Herons are the past masters of this approach, often standing absolutely stock still for several minutes at a time before plunging their powerful neck downwards to seize a passing fish with that sharp, dagger-shaped bill. Kingfishers also use this approach, generally sitting on a branch protruding across the water, so that they can see any passing fish.

One species of heron, the Black Heron of sub-Saharan Africa, has evolved a remarkable way of solving one particular problem: that of bright sunlight obscuring its vision by causing reflections off the surface of the water. As it feeds, it wraps its wings completely around its head and neck for a moment or two, shading its vision and presumably allowing it to see its prey in the water. Having made a grab for a fish or amphibian, it goes through the same movement again until it has caught enough prey to satisfy itself.

Taking the plunge

Other fish-eating species take a more active approach to feeding, diving directly into the water from above. This strategy is practised by seabirds and waterbirds from several families, although one family, the gannets and boobies, has almost perfected this method. Gannets fish by plunge-diving from several metres above the sea, folding back their wings at the very last instant in order to gain maximum depth. Terns also dive from the air, though with more grace than the sheer power of the Gannet.

Osprey is another species which fishes from the air, though using a different approach. Ospreys – or 'fish-hawks' as they were once known – fly along horizontally before dipping down to plunge their talons into the water, grab the fish (using specially adapted toes which can point two forward, two back) and fly off.

Most kingfisher species, as has been mentioned, dive from a perch, but one, the Pied Kingfisher of Africa and Asia, hovers in the manner of a Kestrel, flapping its black-and-white wings like a giant moth before diving down into the water beneath. It holds an unusual record, being one of the largest birds able to hover motionless in the air.

GETTING HELP

Parasitism – the system by which one organism lives off another – is common in nature; symbiosis, in which organisms gain mutual benefit, is less frequent, though it is still found in many animal groups. But a third way in which organisms interact, commensalism, is not only relatively infrequent, but also less well known.

Commensalism basically describes the situation wherein one species benefits from another, with the second species neither gaining benefits nor drawbacks from its association – indeed, usually remaining supremely indifferent. Often this involves a smaller species taking advantage of a larger one, such as the many passerines that nest in the structures of larger species, like House and Tree Sparrows in White Storks' nests. Cattle Egrets are, as their name suggests, commensal with cattle and other large grazing mammals; the trampling of the cows' feet not only digs up invertebrate food, but the dung from the cattle also attracts flies and other insects.

Taken for a ride

Two famous examples of commensalism are the feeding habits of European Robins and a pair of African species, the oxpeckers. The latter are well known for their habit of riding on the backs of large grazing mammals such as buffalo, antelopes and zebras, as well as on domestic livestock such as cattle. By riding on the animals' backs, the birds not only get a good view around them (thus avoiding predators), but they also gain protection – few predators are going to attack! But the main advantage is again related to food: oxpeckers feed on the insects that perch on the mammal's hide, and thereby possibly give minor comfort to the animal itself.

Robins were originally shy, woodland birds, and learned to follow wild boar as they dug up the leaf-litter to find truffles and other food. By following close behind,

OPPOSITE ABOVE: Though supremely adapted for catching fish, Ospreys only succeed once in every five dives or so.

OPPOSITE BELOW: Oxpeckers have developed an extraordinary relationship with large mammals on the African plains.

Robins took advantage of the various worms and insects disturbed by the boars' feet. Today, of course, wild boars are a rare sight in Britain, particularly in the garden habitat favoured by many breeding Robins, and so the birds have learned to follow human gardeners instead!

AERIAL PIRATES: KLEPTOPARASITISM

The act of theft has a long pedigree, especially among birds. Whenever the opportunity arises for one bird to snatch the food of another, it usually will; whether the situation involves a brood of siblings in a nest or birds from the same species in a flock. Birds often steal from other species, too. Gulls are especially adept at snatching bread thrown by children for the local ducks, while in general the larger, tougher birds will tend to dominate in any feeding situation.

But there is one kind of specialized thieving which is so advanced it gets its own technical name: kleptoparasitism. Kleptoparasitism is really just a technical term for piracy: the deliberate and persistent chasing of another bird until it drops its catch and allows the raider to snatch its ill-gotten gains. Generally, like most forms of bullying, it involves a larger, stronger bird chasing a smaller, weaker one: for example, when large birds of prey such as White-tailed or Bald Eagles chase and harry an Osprey and make it drop its newly caught fish.

High sea piracy

Not all groups of birds practise piracy: it is unknown (or at least unrecorded) in parrots, gamebirds or pigeons and doves, and appears to be very rare among songbirds. However, it has been taken to new levels of skill and finesse by two groups of seabirds: frigatebirds and skuas. Frigatebirds are

BELOW: Some birds, like this hungry and opportunistic cormorant, will grab food wherever they can!

OPPOSITE: Great Skuas are well known for their piratical habits, as in this twin-pronged attack on a helpless Gannet.

large, slender and long-winged creatures whose very air has a piratical look about it. Their victims are usually boobies and tropicbirds; which in the case of the former are too unwieldy to escape, and in the case of the latter, too delicate to fight back. Frigatebirds are so acrobatic that they can often catch the regurgitated or dropped food before it even hits the surface of the sea below!

Skuas (known in North America as jaegers, after the German word for hunter) are also very acrobatic and streamlined birds, with great power and strength in their flight. They usually harass terns or kittiwakes, and can make them drop food almost immediately (though some chases go on a little longer as the unfortunate victim attempts to escape).

HUMAN INFLUENCE ON FEEDING BIRDS

In the past century or so, the influence of human beings on the way birds feed has been considerable. Indeed, it could be argued that directly or indirectly, this has had a greater effect on birds' daily lives and habits than almost any other factor.

Human influence on the feeding habits of birds falls into two categories: direct and indirect. In the indirect category, the way we farm is probably top of the list, with both a negative effect on bird populations (the declines due to the extensive use of pesticides, herbicides and insecticides) and, in some cases, a positive one: by changing to certain crops or livestock, or in recent years by going organic, we may benefit birds.

Landfill birdfood

Farming aside, the way we dispose of unwanted food has changed dramatically. At least until the Second World War, and probably until the consumer boom of the 1960s, we rarely wasted food: it was either eaten or saved for another day. Then, as western societies became more affluent, they became more wasteful, throwing away vast quantities of food that, while perhaps not very appetizing for us, was perfectly edible by birds' standards.

At about the same time, legislation was passed in Britain preventing the burning of household waste, in order to reduce air pollution. Unfortunately, as so often happens, the benefits were displaced by unexpected drawbacks; notably that we now have a problem with where to put the ever- increasing amounts of waste we are generating. By dumping it on open landfill sites we have encouraged large populations of gulls and crows to move into populated areas, where they now have a source of food which, in the gulls' case, is far more reliable and abundant than that by the seaside, where they used to live. They have even found places to nest – the flat roofs of our city-centre buildings, which are ideal substitutes for cliff faces!

Garden gains

But the one way in which our influence has really changed the daily lives of birds is a far more direct one: by choosing to feed birds in our gardens. A century – even fifty years – ago this was almost unknown: waste food simply did not exist. Then, in line with our greater affluence, and fuelled by our growing concern for the welfare of wild creatures, feeding garden birds became very popular. Today, the majority of householders in Britain feed the birds, spending considerable sums of money on bird-food products. But compared with North America, we are mere amateurs: the 250 million or so citizens of the United States spend an estimated $2.2 billion every year on feeding birds; more than the Gross National Product of many of the world's poorest countries!

How has this affected the birds themselves? Originally, of course, we fed them in the winter months, especially when hard weather meant that natural food was hard to find. Today we tend to feed them all year round, enabling some birds to become highly dependent on our generosity. Whether this has an effect on bird populations as a whole remains unknown, but we do know that it gives millions of people many hours of pleasure.

OPPOSITE: Several species of gull have recently moved inland to take advantage of the food available at landfill sites.

BELOW: Goldfinches are one of many species that regularly come to artificial feeders in gardens.

Predators and prey

One major side effect of providing a free lunch for the birds in your garden is that you may also be inadvertently providing a free lunch for some of their predators. Predation is a contentious topic, with many people believing that an increasing number of predators will result in a fall in bird numbers. However the population of a predator can only increase if that of its prey is also increasing; if a predator's favourite food starts to disappear, it will also decline.

SONGBIRDS, SPARROWHAWKS AND CATS

Seeing a Sparrowhawk or domestic cat snatching a bird can be understandably distressing, but may not make much difference in the long run. Indeed, there are many myths surrounding this subject, some of which are worth clearing up.

First, it is important to remember that however many songbirds are taken by predators in an area, there are plenty more where they came from. This may sound cruel or heartless, but it includes an important truth – that the vast majority of small birds hatched in a particular year will not survive. Even if they do survive, their life-

OPPOSITE: Every year, many millions of garden birds fall victim to the claws and teeth of the domestic cat.

ABOVE: The Sparrowhawk is one of the most fearsome and effective of all garden bird predators.

spans are extremely short – usually only three or four years at most, often much less. If all birds survived, there would be masses of them within a couple of years; predation is just one of the many causes of death (others include starvation, accidents and disease).

Thus the notion – often given full rein in the popular press – that the decline in some songbird species is a direct result of the increase in Sparrowhawks simply has no biological basis: if all the songbirds went extinct, so would the Sparrowhawk.

The same is not quite true, however, in the case of predation by domestic cats. The problem here is that we have changed the situation drastically: not only have we introduced a predator against which small birds have little or no natural defence, but we continue to support and feed cats even when they fail to catch any prey themselves. Therefore we have given cats an unfair advantage in the battle between songbirds and their predators.

RAPTORS

The very word 'raptor' suggests the sheer power of some of our most admired birds. Its definition is a difficult one, though: if we take it to mean birds that eat meat, it includes many species never considered in the raptor category, such as Blackbirds; after all, they eat worms! Even if we confine it to birds that hunt, kill and eat other animals, then there are many exceptions to the rule, such as herons, which frequently take small rodents or other birds. And what about owls, which are equipped with all the usual attributes we associate with raptors: a hooked beak, sharp talons and the ability to hunt and kill their prey?

In fact, the term raptor – essentially

synonymous with 'bird of prey' – is usually used for the members of the two orders Accipitriformes and Falconiformes. The former, larger, category comprises about 250 species in several groups, including eagles, hawks, buzzards, kites, harriers, Old World vultures and the fish-eating Osprey. The second, smaller, category includes just over 60 species, the falcons and caracaras. Both groups are found throughout the world (apart from Antarctica), and have successfully colonized many different ecological niches, due to their ability to exploit available prey animals.

Food for falcons

Take the falcons. Many are found in the temperate areas of the Old and New Worlds, where they prey mainly on small rodents such as voles and mice, although some species have adapted to take flying insects (the Hobby and Red-footed Falcon) or small migrant birds (Eleonora's Falcon). Indeed, Eleonora's Falcon has so adapted its lifestyle to the availability of food that its young are still in the nest as late as September, thus coinciding with the maximum number of small migrant birds crossing the Mediterranean area, where it lives.

The larger falcons have also adapted to very different extremes of climate: Gyr Falcon, the world's biggest species, lives mainly around the Arctic, where it preys on a range of food. This includes Ptarmigan as well as, in some areas, ground squirrels and that quintessential Arctic species, the lemming.

In contrast, the Barbary Falcon, a paler,

RIGHT: The Peregrine is one of the most adaptable birds of prey globally, able to cope with life in many different habitats. This makes it one of the world's most cosmopolitan birds.

ABOVE: The African Fish Eagle is armed with huge talons, ideal for seizing and killing fish.

smaller version of its close relative the Peregrine, lives in the desert regions of North Africa and Asia, where it hunts mainly birds, including sandgrouse. Meanwhile, the Peregrine itself is one of the most widespread and cosmopolitan species, found in the Americas and throughout much of the Old World.

Surprising methods

Other birds of prey have specially adapted ways of feeding: Marsh, Hen and Montagu's Harriers fly low over their reedbed, moorland or farmland habitat respectively, looking for small mammals or birds, and then using an element of surprise to plunge down upon their unwary prey.

Red Kites tend to scavenge, or even take worms; whereas Black Kites have turned into more or less full-time scavengers on the garbage tips and landfill sites of the world, hence their widespread distribution. The various races (or perhaps full species) in the 'Black Kite' category include the Yellow-billed Kite of sub-Saharan Africa and the Pariah Kite of India; together they are the most successful and numerous raptor in the world.

Eagles have an image as majestic killers, but they too often scavenge for food. When required, however, they can kill hares, grouse and other medium-sized prey with their powerful talons. Incidentally, although eagles and other raptors have a large, hooked bill, this is not generally used to catch or kill prey (which is done using their feet); rather, it is employed after the creature is dead, to tear off strips of flesh on which to feed.

Specialized feeders

In order to exploit all of the possible foods available, birds have adapted their structure, senses and behaviour. The pressures of competition mean that some go to extreme lengths to succeed. Owls feed at night, hummingbirds hover, shrikes keep a larder and some kingfishers avoid water!

OWLS

Another group which sometimes comes under the definition of 'birds of prey' is the owls. The world's two hundred or so species of owl are certainly among our most mythologized birds. This is due to two main factors: firstly, unlike the vast majority of birds, they have forward-facing eyes, giving them a quasi-human appearance; and secondly, they tend to hunt (and indeed carry out most of their activities) mainly at night.

BELOW: Owls such as this Barn Owl have extraordinarily good night vision, which enables them to hunt in the dark.

Night vision

Owls have adapted to hunting and feeding by night in a number of ways. They have proverbially good eyesight, although their night vision is not quite as good as has been claimed in the past. They tend to have large eyes – in some cases even larger than our own – and because their eyes face forward, they have some degree of binocular view lacking in most other birds. They also retain a very wide field of view, an essential tool in initially locating their prey.

Their eyes do differ from other more diurnal species, however; for example, in having more rods (light-gathering cells

particularly good at low light levels) than cones (cells better at seeing colour, and during brighter light levels). Thus they may not be as good at seeing different colours as other birds, but make up for this by having much better night vision, when the ability to distinguish shapes and objects is more important than telling their colour.

Hearing aids

Owls living in open areas, such as the Barn Owl, tend to use hearing rather than sight to locate their prey. They do so by means of their heart-shaped face, which creates a disc that serves to concentrate and magnify sound such as the distant rustling of a field vole. Like other owls, Barn Owls have a virtually silent flight, thanks to their very soft, downy feathers; these enable them to hear even the slightest sound of their prey, home in on it and then pounce. In the Arctic, Great Grey Owls have such sharp hearing that they are able to hunt and catch lemmings even as they tunnel beneath the snow!

Like the day-flying raptors, owls have adapted to a very wide range of habitats and climate types, being found everywhere from within the Arctic Circle (Snowy, Great Grey and Northern Hawk Owls) to the burning deserts of North Africa (Pharaoh Eagle Owl). As with many creatures, northern species – and more northerly populations of the same species – tend to be larger in size than more southerly ones; an adaptation against the cold known as Bergmann's Rule, after the scientist who first proposed it.

On the menu

We know rather more about the feeding habits of owls than for many families,

BELOW: Short-eared Owls are often seen hunting by day, as they are less nocturnal in their habits than other owls.

OPPOSITE: Red-backed Shrikes regularly impale prey on thorn bushes, hence their alternative name of 'butcher bird'.

Shrikes and larders

The predatory habits of shrikes make them good candidates for inclusion as 'birds of prey' – they are certainly the most raptor-like of all the passerines.

The 30 or so species of true shrike may be small, but they can also be vicious. With their upright stance and hooked bill they look like miniature falcons, and have similar feeding habits. Shrikes eat a variety of prey, including insects, lizards, amphibians and small birds, much of which is caught by sitting in wait on a perch, then dropping down to seize their victim.

But their most surprising behaviour comes when there is an abundance of food. It is then that shrikes create a 'larder', where they can keep it safe until they are ready to eat it. The Red-backed Shrike does so by impaling its victims on the thorns of bushes, forming a grisly spectacle which has led to the species nickname of 'butcher bird'.

Red-backed Shrike, once fairly common and widespread as a breeding bird in southern Britain, is now extinct as a British breeding species, probably as a result of wetter summers bringing a reduction in the large flying insects it eats. However, it is still found nearby in Europe, and there is a chance that climate change may enable it to recolonize Britain in the future.

partly because they are unable to digest the fur, teeth and bones of their prey; instead of excreting these items, they cough them up as balls of fluff known as pellets. These items can be examined by scientists, and the exact prey items (species and number) can often be reconstructed more or less exactly.

By careful observation, we also know that not only do owls eat their expected diet of rodents and small birds, they have also been known to take earthworms (Little Owl), tree-frogs (Seychelles Scops Owl), salamanders (Barred and Eastern Screech Owls), snakes (Great Horned Owls) and bats (many species). Indeed, Great Horned Owls have been observed waiting outside a famous cavern in New Mexico until millions of Mexican freetail bats emerge, then flying through the dense cloud of bats until they get lucky!

HUMMINGBIRDS

The world's 320 or so species of hummingbird are more closely adapted to the pollination cycle of flowers than any other group of birds; indeed, many thousands of tropical plant species in the New World rely entirely on these little birds for pollination. Both plants and hummingbirds benefit, for the birds themselves are attracted to the plants in the first place by the provision of large amounts of energy-rich nectar.

Plants and hummingbirds have each adapted to make the most of this mutually beneficial relationship. The plants tend to be brightly coloured, normally red, orange or yellow (three colours to which birds' eyes are especially sensitive). They tend to have large, ornate blossoms, often with markings that help guide the hummingbirds to the centre, where they can obtain the maximum amount of nectar.

Hungry hoverers

The birds themselves have also evolved several important adaptations in order to make the most of this bounty of nectar. First and foremost is their extraordinary ability not only to hover motionless in mid-air, but also to manoeuvre themselves forwards, sideways and even backwards in

LEFT: Hummingbirds have evolved quite amazing powers of flight in order to feed on the nectar of tropical flowers.

OPPOSITE: This White-throated Bee-eater will kill its prey by bashing it on a handy twig before devouring it.

Bee-eaters

Perhaps the most extraordinary birds in terms of their preferred diet are the aptly-named bee-eaters. Found throughout the tropical, sub-tropical and warm temperate climate zones of southern Europe, Africa, Asia and Australasia (but absent from the New World), the 25 species of bee-eater are among the most beautiful and exotic birds of all. They also feed entirely on flying insects, especially bees.

But how do bee-eaters avoid being stung, for they are not immune to stings, as many people suppose? In fact, they do so by a remarkable habit of rubbing and bashing the rear end of the caught bee against a branch or twig until the stinger is removed; they then swallow the de-venomed insect whole!

Bee-eaters also eat a wide variety of other stinging (and stingless) insects, including wasps, beetles, flying ants, termites and dragonflies – almost all caught in flight on their long and graceful wings. Although insects appear to our eyes to fly fairly fast, most are in fact pretty slow, and the bee-eater is usually able to out-manoeuvre them with a few twists of its wings. Bee-eaters will also sit on the back of large animals such as buffalo, using them both as a lookout post and for their ability to attract plenty of insects.

order to move from blossom to blossom and plant to plant in the most efficient way. This is vital because hovering uses up huge amounts of energy, so the bird needs to maximize the amount of nectar it gets on each feeding foray.

Hummingbirds have also evolved a wide range of bill shapes and sizes. Some, like the jacobins and many smaller hummingbirds, have shortish, stubby bills, ideal for probing into short, stubby blossoms. Others, such as the hermits, often have longer, decurved bills in order to reach the nectar concealed deep in longer, more fluty-shaped blossoms.

Big bill

One species, the extraordinary Sword-billed Hummingbird of the Andes, has a bill even longer than its body: roughly 3½–4 inches (9–11cm), compared to total bill and body length of 6½–9 inches (17–23cm). This has evolved because the flowers it chooses to feed on have extremely long corollas. Incidentally, this species also occasionally hawks for insects, opening its bill wide to catch them!

Because their energy requirements are so high, hummingbirds are also able to slow down their metabolism at night by up to 90 per cent, thus reducing their energy expenditure considerably. The next morning they awake as light and heat enable them to feed once again.

FRUIT-EATING BIRDS

Many birds eat fruit from time to time: windfall apples provide much-needed autumn and winter sustenance to thrushes and blackbirds, while smaller fruit such as berries are consumed by a wide range of birds, including Waxwings, Whitethroats and Mistle Thrushes. But a few birds specialize in fruit-eating – those known as 'frugivorous', or fruit-loving, species.

Of these perhaps the best known and best adapted to a fruit based diet are the toucans. Some might be forgiven for thinking that a toucan's diet consists mainly of a well-known brand of dark beer, but in fact the 34 species in the family mainly eat tropical fruits, found in the jun-

LEFT: The extraordinary bill typical of the members of the toucan family enables them to eat a wide range of tropical fruit.

OPPOSITE: Antbirds have learned to exploit the hunting techniques of army ants in order to catch their prey.

Antbirds

Another South American family of birds may not be quite as brightly coloured or as bizarre in its appearance as the toucans, but in terms of amazing behaviour, it's a close contest.

The antbird family comprises well over 200 species, all found in the jungles of South America, where they often feed in mixed flocks. They do not get their name from actually feeding on ants, as you might expect; but rather from their extraordinary habit of following troops of army ants in order to find food.

As the army ants travel across the jungle floor, all other creatures, from insects to rodents, try to escape their relentless march. By following the ants at a safe distance, the antbirds can then feed on the creatures flushed out of their lairs. Not all species regularly follow army ants, but those that do can expect a regular supply of small amphibians, beetles, spiders and lizards. For the visiting birder in the tropical rainforest, the best way to see these birds is to follow the ants – again at a safe distance – then be prepared to try to identify several dozen species of antbird as they pass by!

gles of their Central and South American home. In doing so, they perform a vital role in spreading the seeds of large fruiting trees to new areas.

Toucans eat a veritable fruit salad of different fruits: mainly those of fruiting trees, and including small berries as well as larger fruits. In captivity, they will eat virtually any fruit offered, from tiny berries to huge watermelons – though not, of course, swallowed whole!

An extraordinary beak

The bill of the toucan is rightly regarded as one of the most extraordinary structures found anywhere in the bird world. It has the perfect combination of lightness and strength, with a honeycomb structure that allows it to grow to a huge size – almost half the bird's total body length. Toucan bills have a serrated edge, enabling them to grip fruit more easily; this takes time to develop, and young birds do not possess it. These are not true teeth, but they do serve a useful purpose.

When feeding, toucans will often defend a particular tree against incoming birds, thus keeping the lion's share for themselves. They also forage in small groups, often dropping to the forest floor in order to pick up any fruit which has fallen there.

Yet these extraordinary birds are not entirely vegetarian; several species will take the opportunity to rob other birds' nests and eat the eggs or young, which do not stand much of a chance when faced with that huge, powerful beak. Others will

ABOVE: Kingfishers are, as their name suggests, one of the ultimate aquatic predators.

OPPOSITE: Skimmers have a longer lower mandible, enabling them to practise their unique feeding technique.

hunt and catch lizards, frogs or insects.

One other bird of the American tropics is also a fruit-eater: the extraordinary Oilbird. This species – related to the nightjars, nighthawks and potoos – is the only nocturnal fruit-eating bird in the world. It lives in remote caves, emerging at dusk to search for food.

KINGFISHERS

Odd though it may seem, not all kingfishers eat fish. Indeed, of the 90 or so species in the family, well over half feed at least sometimes on other prey, while several species hardly eat fish at all. Some do not even live by water, preferring more densely wooded or scrubby habitats, where they forage for food on the ground.

Our own familiar Kingfisher (sometimes called the Common Kingfisher, to distinguish it from its many relatives) does feed exclusively on fish and other aquatic creatures such as insects or tadpoles. It is well adapted to do so: its short, stubby body and powerful bill enable it to dive into water, grab its prey and emerge in one rapid movement; while its short, stubby legs are ideal for perching in wait.

Other close relatives of the Common Kingfisher, such as the Malachite Kingfisher of Africa, or the various dwarf and pygmy kingfishers of Africa and Asia, have similar habits, diet and body structure; while the larger, bulkier kingfishers such as the Belted of North America also take fish.

Battered fish

Kingfishers have evolved a very effective way of dealing with the fish once they have caught it: usually returning to their perch, then bashing it on the branch or twig before swallowing it – always head first. One does occasionally see a kingfisher carrying a fish with its tail inside the bird's mouth; this is invariably a courting gift from a male bird to a female, the gift being offered in a way that makes it easy for her to swallow it. Not all kingfishers return to their perch – Pied Kingfishers on Lake Victoria in Africa have learned to swallow their catch whole in mid-air rather than wasting energy by flying off elsewhere to consume it.

The Australian kookaburras – which are also members of the kingfisher family, but have adapted to an almost exclusively terrestrial existence – do the same with their prey of snakes or lizards. In the case of a snake, this is often crucial, as the reptile not only needs to be killed but de-venomed before it can be safely eaten or taken back to the nest for hungry chicks.

The woodland kingfishers of Africa have a slightly different choice of diet; theirs is largely made up of insects and their larvae, including beetles, locusts, grasshoppers and dragonflies!

Skimmers

For a unique feeding style, it is hard to beat the skimmers. The three species of skimmer are found in North and South America (Black Skimmer), Africa (African Skimmer) and Asia (Indian Skimmer), and all share a unique adaptation. They are the only bird family in which the lower mandible of the bill is longer than the upper one, giving them a peculiar, even comical, appearance.

The reason for this extraordinary bill shape becomes clear as soon as one sees skimmers feeding. Living up to their name, they fly low over the water, dipping the lower half of their bill just below the surface. As soon as the bill detects an item of food, the bird jerks its head down, snapping its bill shut and grabbing its prey.

Drinking

As with all other animals and plants on earth, water is absolutely vital for bird survival. However, not all birds need water as regularly as others; indeed, some, like the proverbial camel, can go a very long time without drinking at all.

HOW DO BIRDS DRINK?

Many birds eat moist food such as insects, which have plenty of water content; insectivorous birds such as warblers, flycatchers and robins rarely have to drink more than once or twice a day. However, seed-eaters such as finches, sparrows and buntings need water more regularly, and will visit ponds, pools and bird baths more often than their insect-eating counterparts, especially during prolonged periods of hot, dry weather.

One of the main reasons why birds do not have to drink as frequently as we and other mammals do is that they do not pass urine. Instead, they reabsorb much of the waste water into their own bodies and then excrete fairly dry faeces, usually containing the bare minimum of fluid.

Birds generally drink only fresh water, but they can drink brackish water when necessary. Seabirds are even more adaptable, and are able to drink salt water because they have a special gland that enables them to excrete surplus salt – essential for ocean-going species that may not touch land for days, weeks or even months on end.

BELOW: Pigeons are able to pump water down their throat with their tongue; this enables them to drink quickly.

Swallowing act

How birds drink is often misunderstood; we expect them, like dogs perhaps, to lap up water. In fact, most birds are unable to do so: many lack a swallowing reflex, and so have to dip their bills into the water and then lean back so that the water runs down their gullet. This makes them very vulnerable to predators, and so many smaller birds will drink in flocks, with some birds always on the lookout for an attacker. Other birds, such as pigeons and doves, are able to use their tongues as a kind of pump, to funnel the water down their throat; this enables them to drink more rapidly and efficiently than others.

Some birds are able to drink in flight, notably swallows and martins. These fly low over the surface of a lake or pond, dipping their heads down as they cross to grab a beakful of water. Ironically, it was this habit which once led to the mistaken (and frankly incredible) belief that these birds hibernated beneath the surface of ponds in winter, rather than (as we now know) migrating south to spend the winter in Africa.

BELOW: Sandgrouse store water in their uniquely adapted breast feathers.

Sandgrouse and desert life

One of the most amazing feats of water retention is that performed by sandgrouse, arguably the family of birds best adapted to life in the desert. The 16 species of sandgrouse are partridge or pigeon-like birds, found in suitable dry habitats throughout Africa and southern Asia.

Sandgrouse are adapted to desert life in many ways: they are well camouflaged and generally hard to see, the one notable exception being when large flocks fly (normally once a day) to find water. Most do so in the early hours of the day, when raptors are least likely to be up and about and looking for a victim. They can often be heard at some distance before the flock arrives; it then circles once or twice to make sure all is safe and clear, then lands – still on the lookout for an ambush.

Interestingly, not all sandgrouse drink every day – indeed, some appear to go several days before they need a drink. They have also discovered an extraordinary way to carry water back to their chicks. While drinking, the male soaks his belly feathers in the water; then flies back to the nest. Once he arrives, he exposes his belly by standing upright in a special posture which attracts the chicks, who then emerge from their hiding place and drink to their hearts' content.

BREEDING

As with other living creatures, for birds the essential purpose of life is to reproduce: to pass on their genes to as many offspring as possible, and ensure the continuation of their genes through succeeding generations. This has given rise to some extraordinary courtship behaviour, perhaps nowhere more so than in the various displays, postures and plumages of birds. This chapter begins with an overview of 'the race to reproduce' and its implications for birds and their behaviour. The 'breeding season' is then studied, beginning with timing and the occupation of territories, and followed by a section dealing with that most delightful form of display, birdsong. A section on visual courtship displays is followed by those on nests, eggs and chicks. A final section looks at unusual forms of breeding behaviour, including brood parasitism (the method favoured by cuckoos) and the strange behaviour known as 'lekking'.

RIGHT: A healthy brood of White Storks is guarded by one parent while the other goes in search of food.

The race to reproduce

Most living things spend their lives fighting to survive so that they can breed and raise as many descendants as they can as quickly as possible. This 'race to reproduce' is ingrained into the behaviour of birds, just as it is with any other animal, and in this case has given us some of the most fascinating aspects of wild behaviour on the planet.

A PEACOCK'S TAIL

Take the peacock – or to give the bird its correct name, the male Indian Peafowl. Peacocks are justly famed for their extraordinary tails (in actual fact their upper tail coverts; though for sake of clarity we shall continue to refer to 'tails'). These are unbelievably long feathers (several metres in some specimens), each studded with iridescent patterns in blue and green that resemble a series of eyes. In daily life, for example when hunting for food, the peacock keeps its tail folded together in a horizontal position behind him. But when a female appears whom he would like to woo, he lifts it into a vertical position, spreads his feathers and produces the extraordinary sight we know and admire.

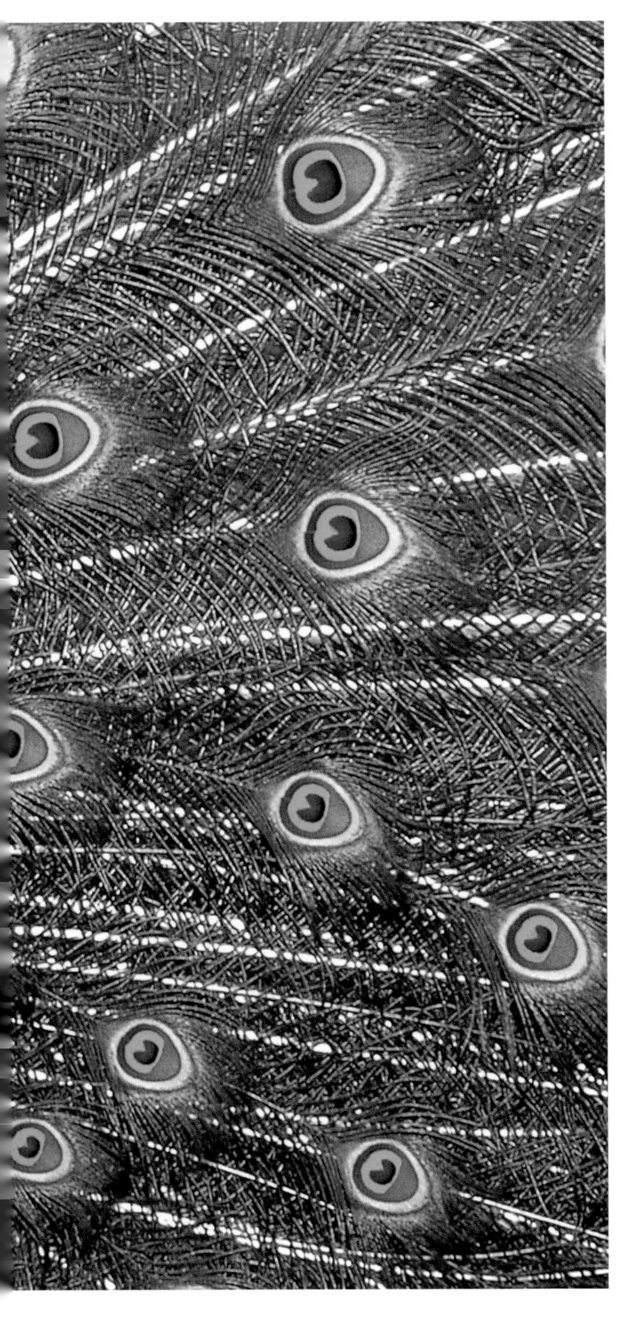

The peacock's tail, along with many other amazing aspects of bird plumage used in courtship displays, evolved in a runaway case of what biologists call 'sexual selection'. Put simply, ancestral males with the longest and showiest tails attracted the fittest females, and then passed on their genetic inheritance to their offspring. This resulted not only in their male descendants having larger and showier tails, but also in their female offspring having a preference for such males.

Puzzling plumage

The success of this strategy used to puzzle scientists, who argued that carrying around such a large tail the whole year round would surely be a handicap to the male birds. Up to a point this is correct, and explains why some birds with spectacular plumage features, such as the African widowbirds and whydas, shed their long tail feathers outside the breeding season. It is also true that had there been a predator able to take advantage of the peacock's limited abilities to make a quick escape, the evolutionary pressure might have eventually led to birds with short, stubby tails, which would give them an advantage when being chased.

LEFT: The male peacock's extraordinary feathers are his major weapon in trying to win the race to reproduce.

Timing

Birds breed at very different times of year, depending on whether they live in equatorial, tropical, temperate or polar regions. This is dictated by what kind of food they give to their young and whether they are resident birds or migrants, which have to make a long journey back to their breeding grounds before they can start raising a family.

WHEN DO BIRDS BREED?

The most important factor deciding when birds begin breeding is food: when the chicks hatch, will there be enough to feed them?

By and large, birds living in the temperate regions of the northern hemisphere, such as Britain, Europe and North America, normally breed during spring and early summer – roughly March to July – in order to hit the peak of natural food supplies such as insects. Residents, especially those which have several broods such as the Blackbird or Song Thrush, are among the earliest starters, often getting going in February (even earlier if there is an unusually mild winter), so that they can raise as many broods as possible during the course of the whole season.

BELOW: Turtle Doves are one of the last summer migrants to return to Britain from Africa, usually arriving in mid-May.

Migrants such as the Barn Swallow or House Martin normally begin breeding as soon as possible after they return (usually in April or early May), in order, as it were, to make up for lost time. Some migrants, such as the Spotted Flycatcher and Turtle Dove, do not normally arrive back until mid-May, and must fit the whole process of staking out a territory, courtship, nest-building, laying, incubation and raising the chicks into a very short period of time, before they head south again later in the summer or early autumn.

Seasonal breeding

For birds living in tropical or equatorial parts of the world, the picture is a bit more complicated. Those living on grasslands such as the African savannah or South American pantanal, which experience seasonal rainfall, normally time their breeding activity to coincide with the wet season, so that there is plenty of food for their young. Such considerations of timing are irrelevant in the vast tropical rainforests of South America, though, as the changes from season to season are so minimal as to be almost meaningless. Here, local factors and microclimates, and their effects on food, will dictate when birds decide to breed.

Summer plenty

Timing is most critical for the few species that choose to nest in the most extreme climates of all: the polar regions. In the Arctic, these are mostly long-distance migrants such as the Red Knot, seabirds such as the Little Auk, or various species of wildfowl, all of which nest there in vast numbers because of the seasonal abundance of food in the brief Arctic summer. These species time their return to their breeding grounds very precisely indeed, often having already paired up and even mated on their wintering grounds farther south. This allows them to get down to laying eggs as soon as they return, maximizing the chances of their young finding enough food to be able to reach adulthood before the need to head back south once again.

BELOW: Crossbills are amongst the earliest breeders of all, often nesting in January and February.

Early birds and late starters

Spring can literally be a moveable feast for birds. For example, the crossbill species found in Europe, Asia and North America often begin breeding as early as January, so that the young hatch when pine seeds are plentiful. Some thrush species, especially Blackbirds, may begin breeding even earlier if the weather is mild. However, this can backfire if a cold spell makes it hard for them to find enough food for their chicks, which will starve as a result.

Seed-eaters, especially buntings, may have young in the nest in July or August, as again this coincides with the maximum availability of seeds. Thus Yellowhammers and Corn Buntings can be heard singing long after most other songbirds have finished.

Territory

A territory is, at its most basic, an area of land defended by a single pair of birds for the purpose of successfully breeding and raising a family. However, this definition is complicated by the different strategies birds adopt to breed, for example nesting in large colonies for safety and protection against predators.

TERRITORIES LARGE AND SMALL

The question 'How big is a typical territory?' is one without a simple answer. For in the bird world, territories can cover any area ranging from hundreds of square miles to a few square feet! As with so many other aspects of breeding behaviour, this is again dependent on the type and availability of food required to feed the hungry offspring.

Thus colonial nesting seabirds, such as auks, gannets or shags, tend to nest very close together indeed, with thousands – sometimes as many as tens or even hundreds of thousands – of birds packed on to a single headland, cliff or island. One reason for this is that they feed on fish, which are found in vast (though sometimes varying) quantities in the seas around where the birds nest. As a result, it is impossible for a seabird to actively defend their food supplies. Another advantage of nesting colonially is that it is a great defence against attack by predators: the

'safety in numbers' rule. By choosing inaccessible places such as offshore islands, many of which are free from mammals such as rats or weasels, seabirds are even safer.

But as a result of limited 'safe space', many seabird colonies are jam-packed with birds, often nesting only a couple of feet apart from each other (or a distance just greater than the pecking range between two birds!).

ABOVE: In order to rear a family, Golden Eagles need a vast territory, sometimes covering several hundred square miles.

OPPOSITE: Seabirds such as Guillemots congregate together in vast colonies in order to breed.

Vast areas

Large birds of prey such as eagles adopt a very different approach. Because their food supply is often scarce and widely spread, and they must hunt over a vast area in order to obtain food, their territories are usually measured in terms of square miles – tens or even hundreds of square miles in some cases. Thus the density of population of a large raptor such as the Golden Eagle will never be very great.

For most birds, the size of the territory will vary, depending on the kind of habitat in which they live, and again on the availability of food. Thus pairs of Blue Tits or Blackbirds nesting in a wood will be less densely packed than those nesting in small suburban gardens, where there is plenty of food (both natural and that provided by humans via bird-tables and feeding stations). As a result, there are often border disputes between males of the same species. In general these are merely 'play-fights', in which one male scares away his rival by means of posturing.

Fight to the death

For one familiar species, the European Robin, things can get much more serious. Despite their benign appearance, Robins are tough little birds, and will often engage in vicious and even violent fights. These often end in injury to the beaten male; occasionally even in his death.

But for the vast majority of species, territories are won and defended by other means: either visual displays by a male patrolling the borders, or (for songbirds, a group which contains more than half the world's bird species) by singing.

Birdsong and calls

Although we find birdsong one of the most delightful of all the sounds of nature, we must remember that the birds are not singing purely for our benefit! Singing enables them to keep a home and a mate. Vocalisation by birds can be quite elaborate in the case of many songbirds, and also quite simple. And a bird's song is a great aid for birdwatchers, allowing them to identify a bird before they have even seen it!

WHY DO BIRDS SING?

Birdsong has two primary purposes: for (usually male) birds to defend a territory throughout the period of time known as the 'breeding season', and for them to win, and then keep, a mate.

Birds defend their territory by patrolling it at regular intervals, often doing a circuit of its boundaries and singing from prominent positions such as the tops of trees or buildings, in order to maximize the carrying potential of the sound. In this way, they hope to keep rival males as far away as possible, though in some habitats such as suburban gardens, territories may only be a few feet apart.

Early in the season, when still unmated, this strategy also has the advantage of bringing the song to the attention of any passing female. She will compare the songs of rival males, usually choosing the one with the most impressive sound: either in terms of volume, variety, persistence or all three.

An early start

In temperate latitudes such as Britain and much of Europe and North America, birdsong usually begins soon after the start of the New Year. In Britain and north-west Europe, the first songsters are members of the thrush family: especially Blackbird, Song Thrush and Mistle Thrush, all of which begin breeding early in the year to maximize their chances of raising several broods in a single season. Other early songsters include the Robin (which often sings throughout the autumn and winter as well), the Wren, and the Dunnock.

All of these species sing most frequently and persistently on fine, clear evenings or in the hours just before and just after dawn, often using promi-

OPPOSITE: The Blackbird has one of the best-known and best-loved of all bird songs: deep, rich and fluty in tone.

ABOVE: The European Robin sings at all seasons, defending territories against would-be intruders.

nent positions such as the roofs of buildings in order to deliver the maximum impact.

Slow build up

Gradually, as spring progresses, other species join in, especially Blue and Great Tits, and members of the finch family such as Chaffinches, Greenfinches and Goldfinches. Meanwhile, during late March, April and early May, the returning migrants begin to arrive, sometimes in a mad rush, brought here on light, southerly winds on a fine spring day. Chiffchaffs and Blackcaps are usually the first to return, followed by Willow Warblers, delivering their delightful, silvery song along woodland edges and on heaths.

By the start of May, the dawn chorus is at its peak, and a visit to a local wood or park in the early hours can be an unforgettable experience. From then onwards, the intensity and volume of birdsong will gradually decrease, as the birds get on with the task of raising a family. By late June and July the same wood may be virtually silent, particularly so on a hot summer's day.

Female vocalists

Although it is the male which usually sings, there are a few exceptions. The best-known is the European Robin; both males and females sing because they defend territories outside the breeding season. In North America, female Red-winged Blackbirds also sing during the breeding season, in order to warn off rival females – male Red-winged Blackbirds are not known for their fidelity and will mate with more than one female if given the chance!

Singing does not always work, however, especially if there is an abundance of unmated males in the area, or if nest-sites and territories are few and far between. In extreme cases, a rival male will even attack the incumbent, engaging in mock-battles or, occasionally, real ones. However, once things have settled down and most birds have paired up for the breeding season, then many species either reduce the time they spend singing or stop altogether. After all, with eggs to brood and hungry chicks to feed, it is at this time that the real work begins.

THE DAWN CHORUS EXPERIENCE

Every spring, strange groups of people across Britain and North America perform an annual ritual, driven by an impulse so primeval that we can be certain that it was shared by our distant prehistoric ancestors. They leave their homes under cover of darkness, gather together in an ancient wood or forest, and wait.

Just before the first light appears as a glimmer in the sky, they hear what they have come to witness: the very first sound of a bird, pouring forth its song in the chilly, pre-dawn air. The dawn chorus experience is about to begin!

Many of those who have witnessed a dawn chorus regard it as a mystical, almost religious experience. They return to their homes or head for a local café for a hearty breakfast, feeling fulfilled and at peace in a way that most people rarely share. For to get the real benefit from a dawn chorus you do need to get up very early in the morning. But for those that make the effort, the reward is immense: a feeling of spiritual oneness with nature, which seems to help put human woes and worries into their correct perspective.

LEFT: The Nightingale has been rightly celebrated by writers, poets and musicians for the beauty and complexity of its extraordinary song.

How do birds sing?

Unlike humans, who vocalize through a combination of the voice-box and larynx, birds produce their songs and calls with the syrinx. Unique to birds, the syrinx enables the bird to produce far more complex sounds than we can, effectively doubling their capacity to produce notes. Like humans, birds are also able to vary their song in four different ways: by changing the pitch up or down; by altering the tone (by making fluty or reedy sounds, for example); by changing the volume; and by altering the rhythm.

Just as human babies learn to speak through a combination of inherited ability and listening to speech, so do birds. A young male Chaffinch, for example, will be born with the basic building blocks of the Chaffinch's song embedded in his brain; but only by listening to the song of his father (and other males nearby) will his song develop. Likewise, female Chaffinches are born with a basic ability to recognize a male's song, and by hearing it as they grow up they learn to judge between a 'good' song (i.e. loud, varied, dynamic) and a 'bad' one (quiet and unvarying).

Experiments with caged birds, which are either exposed to no song at all or played tapes of other species, demonstrate that they produce a poor imitation of their 'proper' song.

Serious singing

Whatever the joys and benefits brought by the dawn chorus, there is one indisputable fact to consider: the birds are not doing it for us to enjoy! What to us sounds like a beautiful orchestra of sound is in fact the frontline in a battle to survive. Spring is the real crunch time for birds: it is the breeding season, when they need reproduce themselves or face the prospect of their genes being extinguished. Especially for songbirds, which rarely live longer than a year or two, this season is the most important time of their life, when the very purpose of their existence is on the line.

For male birds, the urgency is almost palpable. They need to win and then defend a territory, persuade a female (or in some cases, several females) to mate, build a nest, then defend her against any intruders while keeping the usual eye out for danger from predators. Meanwhile the female must choose her mate – carefully, for again her future depends on getting it right – lay eggs, incubate them, then brood her precious young until they are old enough to fledge, leave the nest and begin to fend for themselves. Then, in some species, the pair will start all over again – up to five times in a single season!

Dawn appointment

The prelude to all this frantic activity is the male's song, which may be delivered at intervals throughout the day, but is especially prominent at dawn and, to a lesser extent, at dusk. There are various reasons for this early peak of song activity. First, songbirds find it hard to find food in the dark, so it is better to sing and leave foraging until later in the day. Second, female fertility is at its highest in the early hours of the morning, so the male must make sure that an intruding male does not take advantage of this. Finally, birdsong carries further at dawn: partly because there is less extraneous noise, such as traffic, and partly because sound travels better through still, calm air.

HOW WE LEARN BIRD SOUNDS

Many people struggle to learn bird songs and calls; this is understandable, given their variety and complexity. Some people, especially those with musical talent, find it easier than others, but just like learning to play a musical instrument or, indeed, to speak a foreign language, there is no simple way to master the subject.

Nevertheless, there are a few short-cuts such as the use of mnemonics – helpful words or phrases that enable you to commit a particular sound to memory. For British birds, probably the best known is the Yellowhammer, which is supposed to sing 'a-little-bit-of-bread-and-no-cheeeeese' (with the emphasis on the drawn-out final syllable). The Great Tit's song can be summed up as 'tea-cher, tea-cher' – even though Great Tits are notorious for having dozens of other calls and sounds.

Pied Wagtails say 'chis-ick', while Wood

BELOW: The song of the Chaffinch has an an accelerating cadence which has been likened to a cricket bowler's run-up.

OPPOSITE: The Yellowhammer features prominently in popular country folklore, thanks to its characteristic song pattern.

Pigeons are supposed to intone 'my toe is bleeding' (the emphasis on the third word!). Another bird which is heard far more often than it is seen is the Quail, which allegedly calls 'wet-my-lips'.

Aids to learning

Not all memory aids are strictly mnemonics: others are based on the mental image that the sound of the bird conjures up. These include the likening of the song of the Chaffinch to a cricketer running up to deliver a fast ball, or the Grasshopper Warbler to an angler letting out line from a fishing reel.

Learning these short-cuts is a matter of listening to the song of a bird and waiting until you are struck by a useful analogy which helps you recall it later on. So next time you hear a Reed Bunting, try my personal favourite: to me, it sounds like a bored sound engineer running through a microphone test... 'One, two, testing. One. One. Two. Two. Testing'. Boring perhaps, but effective!

MIMICRY

From time to time, people are puzzled about the origin of a particular sound: perhaps the ringing of a mobile phone, or the sound of a car alarm. For however hard they look, they simply cannot trace the sound's origin. In such cases, nine times out of ten the culprit is the humble Starling, an often overlooked bird with the extraordinary ability to mimic a wide range of natural and human-made sounds. They are quick on the uptake, too, able to learn

Songs and calls

I'm often asked what the difference is between a bird's song and its call. The answer isn't easy, as the song of one species may be highly complex, while its relative has a simple, repetitive song that is more like the call of another species.

Generally speaking, songs are used (mostly by male birds) to defend a territory and attract a mate, usually during the breeding season, while calls are relatively brief and less complex vocalizations, given by both males and females throughout the year. Calls have a variety of purposes; indeed, one species may use different calls in different circumstances, such as sounding the alarm against a predator, for example.

a new sound within days – sometimes even hours – of hearing it for the first time.

Starlings are not the only mimics in the bird world, though they are among the most accomplished; as are their exotic relatives. the mynas. Originating from the tropical forests of South-East Asia, mynas are often kept as pets due to their famed ability to copy not only artificial sounds but also the patterns of human speech. Parrots are also great mimics, of course, though oddly they do not do so in the wild, but have to be taught the skill.

Famous impersonators

Other famous mimics include the mockingbirds of North and South America (aptly named for their extraordinary vocal abilities); the robin-chats of Africa, and in particular the Superb Lyrebird of Australia,

BELOW: The Superb Lyrebird is one the most accomplished mimics, capable of copying both natural and artificial sounds.

OPPOSITE: The Marsh Warbler may not look all that exciting, but it is one of the best mimics in the bird world.

one of which was given star billing on worldwide television when it was shown in David Attenborough's series *The Life of Birds*, impersonating a camera motor-drive and even a chainsaw!

So why do some species have the gift of mimicking other birds, human speech and natural sounds, while most do not? Rather like the peacock's tail, the ability to imitate almost certainly arose as a result of runaway sexual selection, in which those males who were the most accomplished mimics found it easier to attract females. Thus the ability to create a variety of unusual sounds was passed onto the next generation, until it had spread through the whole population.

MARSH WARBLER: TOP MIMIC

Of all the world's birds, which would win the prize for the most extraordinary mimic? It is not, as you might expect, a parrot, nor is it the European Starling. No, the title of global bird mimic extraordinaire goes to a little-known and rather dull-looking member of the Old World warbler family: the Marsh Warbler.

Marsh Warblers are found in the temperate latitudes of Europe and western Asia, with a tiny number still occasionally breeding in south-east England, following the extinction in the late 20th century of the small population in the south-west Midlands. In Europe, the species is usually found on the edge of damp areas, often in dense vegetation, making it a challenge to see. Indeed, the usual indication that a Marsh Warbler is present is its song which, although obviously coming from a warbler of some sort, may include brief extracts from several dozen other bird species!

African repertoire

What makes the Marsh Warbler's song even more extraordinary is not simply the number of different species it has been known to imitate – at least 212 in all – but the fact that these include sounds learned on its African winter-quarters as well as on its European breeding grounds. Indeed, the majority of different songs – a total of 113 – belong to African birds, which is very confusing for the novice birder in Europe!

It is thought that Marsh Warblers learn almost their entire song vocabulary in the first few months of life, and that by the time they have returned to Europe to breed at the age of one year old, they can imitate an average of 77 different species.

Duetting

Not all birds are solo artists: some prefer duets. The phenomenon of singing with a mate is more usually found among tropical species, for the simple reason that there is so much competition – from other kinds of bird as well as fellow males – that duetting is the way for some males and females to keep in touch.

Birds that regularly duet include the African Robin-chat; the male and female respond so rapidly to the other's call that the result sounds like a single song. Various species of African shrike, including the Tropical Boubou, also take part in perfectly timed duets.

But the most bizarre duet comes from the Eastern Whipbird of Australia. The male sings, emitting a pure whistle followed by an explosive sound, like the cracking of a whip. Immediately the female responds with a two-syllable note, completing the duet.

The male creates his extraordinary whip-like sound by making his whistle ascend or descend very rapidly in pitch, from as low as 500 megahertz to as high as 8,000 megahertz, in just one-tenth of a second. It is thought that the actual crack is produced by the sound reflecting off the surrounding foliage. It seems, however, that the female may not be so impressed: recent research has revealed that some female whipbirds will duet with males other than their mate!

Other vocal communication

It is well known that bats have the extraordinary ability of using a navigational aid known as echolocation to find their way around in the dark, avoid bumping into the walls of their caves and locate their flying insect prey. What is much less well known is that some birds have this ability, too.

The two main groups that can use echolocation are the cave swiftlets of South-East Asia, and a peculiar relative of the nightjars, the Oilbird of South America, both of which live in dark caves.

In bats, echolocation works whereby the animal emits a series of very high-pitched clicks (too high for the human ear to hear), which then bounce off any objects in the vicinity and send an echo back to the bat's brain, enabling it to process a mental picture of its surroundings. Bird echolocation is not as complex, nor quite so effective. For a start, the sounds are emitted at normal frequencies, so that humans are able to hear them. They are also not nearly as rapid as those of bats, and are mainly used to get a sense of the space within the caves in which the birds live.

OPPOSITE: The male and female Eastern Whipbirds perform extraordinary duets as part of their courtship display.
RIGHT: The enigmatic Oilbird uses the technique of echolocation to find its way around dark caves.

The evolutionary advantage of such an extensive vocabulary is not obvious, but it is thought that it may be to confuse birds of other species and discourage them from nesting nearby. An alternative reason is that, just like the peacock's tail, it is a runaway example of sexual selection, in which the best mimic gets the best females, and thus passes on the ability to imitate to its offspring.

NON-VOCAL COMMUNICATION

Not all sounds made by birds are vocal. Many species have developed an amazing range of non-vocal sounds, producing a veritable cacophony of 'clicks', 'whirrs' and 'clatters'. Take, for example, the Snipe. During courtship display on their marshy territory, male Snipe perform an extraordinary display in which they fly high in the sky, then fold their wings and plummet downwards. While doing so, they fan their tails, letting the air pass through specially adapted feathers. As it does so, the stiffness of the feathers allows the tail to vibrate, producing a sound rather like a deep wind instrument and known as 'drumming'. This sound can carry long

ABOVE: White Storks frequently shake their heads and clatter their bills noisily in order to display to their mate.

OPPOSITE: Cranes are world-renowned for their wonderful dancing displays. These are Red-crowned Cranes from Asia.

distances and enables the male Snipe to defend his territory and win a mate.

One South American songbird has also developed an extraordinary non-vocal display. Male White-bearded Manakins gather in 'leks' (see page 139), in which several males display together to try to impress watching females. They do so by 'clicking' their wings, expelling trapped air to produce an explosive snapping sound.

Headbangers

Another, often overlooked, form of non-vocal communication is the drumming of woodpeckers. Again, this has the dual function of proclaiming the male's presence in a territory and attracting a female. Drumming may have evolved from woodpeckers' habit of tapping on hollow branches and tree-trunks in order to extract insect food. Whatever the origins, the world's 215 or so species of woodpecker have evolved special features in order to drum without getting a headache: their beak is cushioned by soft, flexible tissue which minimizes the effect of the vibrations, while their neck muscles are specially developed to allow them to drum up to 20 times a second.

Finally, several species use 'bill-clattering' during their courtship displays. Of these, the best known is the White Stork, males and females of which face each other while standing on their huge nest and clatter their bills in what appears to us to be an expression of greeting, perhaps even love.

Courtship and displays

The courtship displays of birds are among the most visually exciting examples of animal behaviour on earth. Their variety and inventiveness takes one's breath away: some are brief and direct, others bizarre and lengthy. But all have a single purpose: for one bird (usually the male) to persuade the other (usually the female, or a group of females) to pair up with him, mate and produce a family.

VISUAL DISPLAYS

When it comes to listing the variety of courtship displays found among the world's birds, it is hard to know where to begin. There are the birds-of-paradise of New Guinea, whose males sport exotic plumes of feathers which they use in ever more inventive ways; some turn themselves upside-down while hanging onto a twig in order to shake their feathers and impress the watching female.

In New Guinea and Australia, the bowerbirds take a more subtle, less direct, approach. Instead of using exotic plumage to attract a mate, they build a 'bower' out of grass and other plant material, then decorate it with a variety of objects – often including discarded man-made artefacts such as the tops of soft drink bottles or an item of jewellery. In such cases, the shinier and more glittering the better. Some members of the family, such as the Satin Bowerbird, actually use a twig to 'paint' their bower using a mixture of natural plant pigments and saliva!

Dancing with delight

Other birds use a more visual, dynamic form of display to attract the opposite sex. Several species of crane indulge in what can only be described as a courtship dance, the most famous of which takes

place at Hornborga in southern Sweden for several weeks each April. Tens of thousands of cranes, newly returned from their winter-quarters in southern Europe, stand in front of each other like competitors in a ballroom dancing competition, leaping into the air in turn while uttering their characteristic honking calls. This has long been celebrated by the local people as a sign that spring has finally returned after the long, harsh, northern winter. Today the cranes are a major tourist attraction, as people come from all over the world to witness their delightful courtship dance.

Making a pass

Many raptors, especially larger species such as eagles and harriers, take part in 'food-passing' displays, in which the male shows his devotion to the female by catching an item of prey, then flying high in the air before dropping it for her to catch in her talons: a spectacular piece of co-ordination by any standards! Other raptors engage in dramatic 'talon-grappling' displays, turning over and over in the air as they plummet towards the ground.

As with much bird behaviour, however, the most bizarre and spectacular courtship displays are often found in the most far-flung places, such as the Amazonian rainforest of South America. The density of species here is greater than anywhere else in the world, and as a result birds have evolved ever more unusual displays. Of all of these, that of the two species of cock-of-the-rock must take the prize. Males gather at display areas known as 'leks' (see page 139) and compete with each other for the watching females, via a fantastical array of dancing, fighting and posturing.

THE DANCE OF THE GREAT CRESTED GREBE

In the summer of 1912, a young scientist named Julian Huxley decided to spend his fortnight's break on what used to be called a 'busman's holiday' – pursuing his passion for birds. In those days, however, not only was birdwatching in its infancy as a hobby, but the study of ornithology was very far removed from the respected science it is today. Carried out mostly in dusty museums, it centred on the dissection and close analysis of dead specimens; the study of living birds and their behaviour, which we so take for granted today, was considered irrelevant.

Nevertheless, Huxley was determined to find out what birds actually did on a day-to-day basis. He chose as his study area Tring Reservoirs in Hertfordshire; and as his subject, several pairs of Great Crested Grebes.

During his two-week stay, Huxley simply noted down what the birds

Reverse sexual dimorphism

Generally, in nature, the male takes the lead in courtship and tends to be the showier of the two sexes. But with birds this is not always the case. In a few species the female not only dominates the courtship ritual and leaves the male to the job of incubating the eggs and rearing the chicks, but she also has a brighter plumage than her mate.

This phenomenon is known as 'reverse sexual dimorphism' and is most easily observed in four species of wader: the Dotterel, and the three phalaropes – Red-necked, Grey or Red and Wilson's. In each case, the female takes over every aspect normally associated with the male, apart, of course, from actually laying eggs!

Red-necked Phalaropes return to their Arctic breeding grounds in May and the females immediately chose a suitable male. Once paired up, the female lays the eggs and then leaves almost all the work to her dowdier mate.

Dotterels do the same, but with a fascinating twist. In the small population nesting in the Scottish Highlands, once the female has chosen a mate and laid her eggs, she will sometimes head off to Norway, find a new mate and breed a second time. Meanwhile, the male is left to incubate the eggs and raise the young on his own.

did, however apparently trivial or irrelevant it seemed, in his voluminous notebooks. It was the middle of the nesting season, and he witnessed a remarkable variety of different types of courtship and breeding behaviour: from the legendary 'penguin dance' performed when male and female grebes face each other in the water, 'stand up' in the air and offer each other 'presents' of water-weed; to the sight of the parent grebes carrying their stripy young on their backs.

Later on, Julian Huxley became one of the 20th century's most respected and best-known scientists, working tirelessly to both conserve the world's birds and to popularise them among the general public. But perhaps his greatest achievement was when, as a young scientist, he opened people's eyes to the extraordinary breeding behaviour of a common and familiar, but little-studied, bird: the Great Crested Grebe.

OPPOSITE: Great Crested Grebes display to one another using water weed in the so-called 'penguin dance'.

BELOW: The female Dotterel is not only brighter than the male, but she also takes the lead in courtship.

Nests

It would be an exaggeration to claim that there are almost as many different kinds of nest as there are different types of bird. But from the hanging nests created by Penduline Tits and the various species of African weaver bird, to the beautiful domed structure built by the Long-tailed Tit (using lichens, spiders' webs and feathers), it is certainly true that the ingenuity and variety of different nest designs is one of the true wonders of the bird world.

DIFFERENT TYPES OF NEST

Imagine a bird's nest and the image is of a cup-shaped structure as favoured by many songbirds such as Blackbirds or Robins. But what about holes in trees, favoured by many woodland species such as woodpeckers or tits? Or floating nests, such as those made by coots and grebes? Or the magnificent structures built by large birds of prey? Or even the minimalist strategy of laying your egg in a slight hollow on the ground, as do many waders and cliff nesting birds such as Guillemots and Razorbills?

Contrasting cousins

The fascinating thing about nests is that even closely related species often adopt a totally different nesting strategy from their relatives. Take the three species of hirundine that regularly breed in Britain: the Barn Swallow, House Martin and Sand Martin. The Barn Swallow, as its name suggests, often nests inside a barn or other farm building, and creates a simple nest from mud, lined with grass and feathers, and usually situated on a wooden beam or in a crevice in a stone wall.

The House Martin also lives up to its

Going underground

Owls and parrots usually nest in holes in trees, where their eggs and chicks are safe from attack by predators, but what can they do if there are few or no suitable trees? In the case of the native owl and parrot found in Patagonia, the answer is simple: go underground!

The Burrowing Parrot is the only parrot to nest underground, digging holes in the sides of cliffs using its specially adapted claws and usually nesting in large colonies. Instead of trees it also uses telegraph wires and posts as places to gather and utter its harsh call. As a result this is the only species of parrot found across a very large part of southern South America.

The Burrowing Owl is much more widespread, being found from the southern United States all the way to Patagonia. Like the parrot it also digs its own holes, and again it has specially adapted feet and claws in order to do so. Burrowing Owls have particularly long legs, enabling them to gain a lookout when perched on the ground.

But perhaps the most extraordinary hole-nester of all is the Magellanic Penguin. This medium-sized penguin nests in huge colonies along the Argentinian coast, with nesting birds concealing their eggs in rough hollows out of the reach of predators such as caracaras and skuas.

name, but in this case creates a nest made entirely from tiny balls of mud which it sticks together under the eaves of houses; a long and painstaking task which can end in disaster if the nest falls down. In the days before houses, this species would have nested on cliffs and crags; indeed, there are still a few colonies of cliff-nesting House Martins around our coasts.

The smallest of this trio, the Sand Martin, adopts a completely different strategy: digging a hole in the sandy banks of rivers (hence its North American name, Bank Swallow).

OPPOSITE: The Penduline Tit builds an extraordinary hanging nest, from which the species gets its name.

RIGHT: The Burrowing Owl nests in underground burrows, which it excavates using its powerful legs and feet.

Mrs Hornbill

Hornbills are found in the tropical areas of the Old World. Most species nest in a hole in a tree or the face of a cliff, and most also follow the bizarre practice of 'walling up' the female inside the nest in order to keep her safe from predators.

This process begins with the female and male making the natural hole smaller by applying large amounts of mud. When she can still just about squeeze in, she enters for the final time, then allows the male to pass leaves and bark to line the nest. Finally, the male gathers more mud and passes it through the gap to the female, who carries out the last stage of the process: imprisoning herself for up to four months!

Inside the nest she is kept cool by a system of ventilation shafts, and also takes advantage of her incarceration to moult her wing and tail feathers. As a result she is completely dependent on the male for food – if he dies she will starve. However, the risk is easily outweighed by the greater safety against predators.

Finally, once the chicks are grown enough to leave the nest, the female breaks down the mud walls and they all emerge – a process that can take several days.

LEFT: Using mud, male hornbills wall up their mate in her nest. They then feed her via a small opening in the 'wall'.

Safety in numbers

Like Sand Martins, many species nest in large colonies, where they gain the benefit from the 'safety in numbers' principle. Seabirds in particular are mainly colonial nesters, and often hardly bother to build a 'nest' at all; simply laying their egg or eggs in a suitable depression in order to incubate them. This is because unlike birds that nest on their own, there is less need to conceal the eggs from hungry predators. Instead, seabirds rely on their collective ability to warn each other of an approaching predator; then if necessary take action by attacking the intruder and chasing it off. Nevertheless, egg and chick loss at some colonies is very high indeed; and some seabirds – notably Puffins – have adopted a safer strategy of laying their eggs underground in rabbit burrows!

Eggs

Thanks to our ability to exploit the Red Junglefowl of Asia, otherwise known as the domestic chicken, the egg is a very familiar object in our daily lives. But eggs come in a variety of shapes, colours and sizes and are usually incubated by body heat, provided by the parents.

EGG SHAPED

We tend to assume that all eggs are 'egg shaped' – oval, with one end larger than the other – whereas in fact eggs come in many different shapes and sizes.

For example, the eggs of hole-nesters such as kingfishers and owls tend to be much rounder – sometimes almost spherical; as are the eggs of the exotic Ross's Turaco of equatorial Africa. In contrast, Great Crested Grebe eggs are long and slender, tapering to a point at either end. The eggs of some cliff nesting auks such as guillemots are 'pear-shaped': tapering to a point at one end and bulbous at the other, probably because this shape tends to roll in a tight circle rather than a wide one. This helps minimize the risk of it falling off a narrow ledge if knocked accidentally by a parent.

Eggs also vary dramatically in size. As you might expect, the larger the bird, the larger the egg. So the Ostrich's egg is about 6½ inches (16 cm) long and weighs up to 3.3 lbs (1.5 kg), about two dozen times the size of a hen's egg. However, relative to its size the Ostrich's egg is in fact one of the smallest, representing between one and two per cent of the adult's body weight. In contrast, the eggs of kiwi species are huge, comprising about one quarter of the female's body weight. This is because kiwis evolved from much larger birds, and as their body size gradually reduced the size of their eggs remained more or less the same.

BELOW: The Mute Swan's eggs are the largest of any British breeding bird. White when first laid, they soon become stained by detritus in the nest.

ABOVE: The Australian Brush-turkey, a member of the megapode family, builds an incredible mound-like nest.

OPPOSITE: The Emperor Penguin is one of few bird species to breed in Antarctica and does so during the long winter.

INCUBATION

Incubation is, at its simplest, the process by which an egg is provided with the heat required for the embryo within to develop and eventually hatch. Normally this heat is provided by one or both of the parent birds, though there are several exceptions to this rule, as we shall see.

Most birds incubate by settling down onto the clutch of eggs and transferring heat from the adult by means of an area of bare skin on the belly known as the brood patch. This patch, which usually has fewer feathers than the rest of the plumage, allows heat to pass more efficiently from the body of the parent into the eggs, and so maintain the optimum temperature.

Shared responsibility

The time spent incubating eggs varies enormously, from an hour or so at a time in the case of many songbirds, to more than two months in the case of the Emperor Penguin (see opposite). Where both parents share the incubation duties, changeovers are usually done as quickly as possible to minimize the time that the eggs are left exposed. Where only one parent incubates, he or she will leave the nest to forage for food, but usually only for short periods. In some cases the non-sitting bird will bring food to the nest to avoid the eggs being left uncovered.

Some species such as grebes cover the eggs with vegetation when away from the nest, though this probably has more to do with avoiding detection by predators than keeping them warm.

The total period spent incubating the eggs also varies considerably, with larger birds tending to spend longer doing so. Passerines generally incubate for between 11 and 19 days, while larger birds of prey and seabirds may incubate for several months.

THE LONGEST NIGHT

When it comes to dedication to parental duty, few of the world's wild creatures can rival the male Emperor Penguin. To start with, Emperor Penguins breed in one of the most hostile environments in the world: the frozen wastes of the Antarctic continent. To make things even harder for themselves, they breed in the depths of the southern winter, which ensures that their single chick reaches the maximum possible age before the onset of the following winter: thus helping to maximize its chances of long-term survival. Worst of all, they breed several hundred miles from the nearest source of food: the open sea.

Like most large, long-lived birds, Emperor Penguins follow the strategy of having a small number of chicks (in their case, one), and have a very long incubation period – an average of 67 days. In fact, this species could not possibly have two chicks for one very simple reason: in the absence of any nesting material, the male penguin incubates the egg by resting it between his feet and keeping it warm by brooding it with his downy feathers. The alternative – laying the egg directly onto the snow or ice – would rapidly end in disaster.

Dark brooding

So as the days begin to shorten and the long, dark Antarctic winter takes its grip, the female lays her single egg, which is rapidly transferred to the male's feet before it can freeze. As incubation begins, the female leaves her mate, walking (for penguins are of course flightless) the three or four hundred miles to the edge of the continent. There, after an incredibly arduous journey, she can find food for herself.

Meanwhile, the male sits patiently, huddling with others against the cold and brooding his single egg. More than two months later, after losing almost half of his total body weight, his mate returns, just as the chick has hatched. He hands over care of their tiny offspring to her; then he begins the long trek back towards the open ocean and food.

Hatching

The incubation of one or more eggs is an achievement in itself: many are taken by predators such as snakes, squirrels or magpies – opportunists always on the look-out for a quick, easy and nutritious meal. So not all eggs make it to the hatching stage and even if they do, the challenge for the bird inside is only just beginning.

The egg – and the embryonic chick inside – 'prepares' for the chick's emergence into the wider world in a number of ways. First, as the hatching date gets near, the eggshell itself begins to weaken and a small sac of air develops at the blunter, more rounded end of the egg. Meanwhile, the chick begins to develop a small bony projection on the tip of its bill, known (erroneously, for it has nothing to do with true teeth) as the 'egg-tooth'.

As the time to emerge from the egg comes, the chick manoeuvres itself up to the blunt end of the egg, and uses its 'egg tooth' to make tiny cracks in the shell. Still manoeuvring itself around the egg, it pushes at various points around the shell until tiny cracks turn into bigger ones. Finally, it is able to emerge; bedraggled, tired, but triumphant.

BELOW: This baby Ostrich breaks out of its egg by using its powerful bill, after an incubation period lasting 42 days.

OPPOSITE: Songbird chicks such as these Blackcaps are born naked, blind and helpless and require constant feeding.

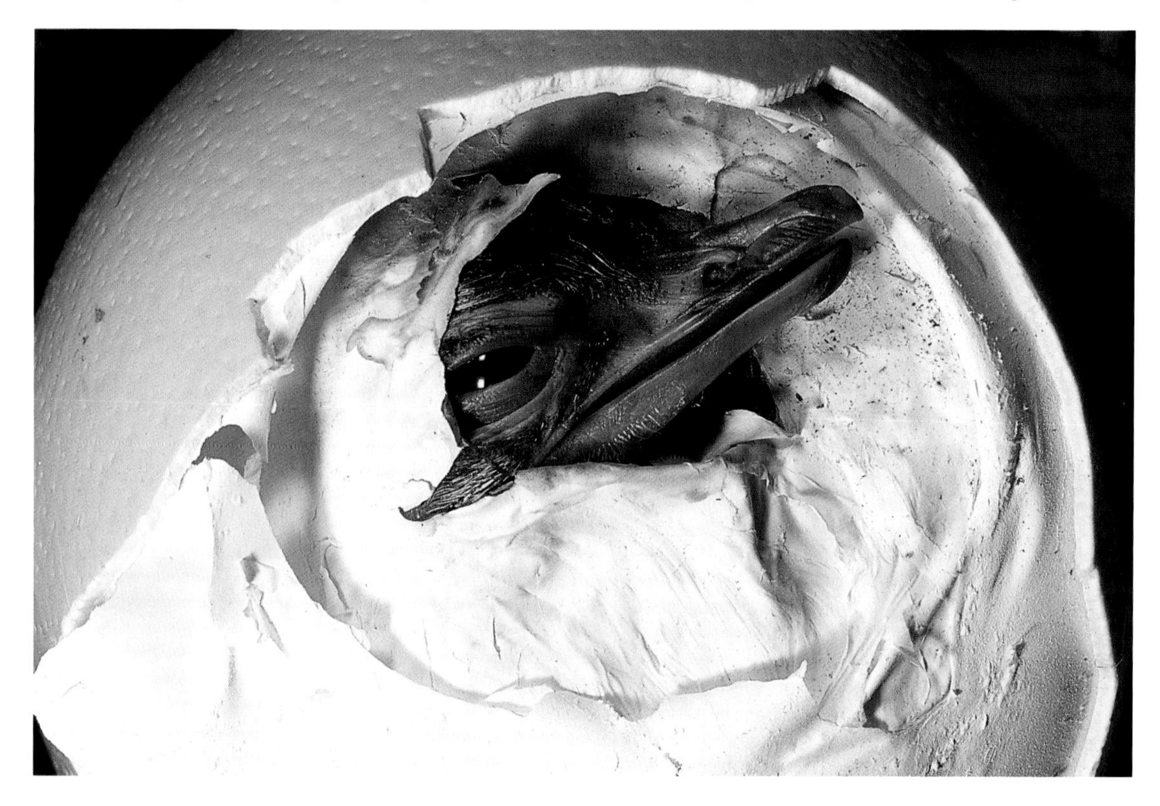

Chicks

Birds follow two basic strategies once the chicks have hatched, though there are many different variations on this dual theme.

SHOULD I STAY OR SHOULD I GO NOW?

Broadly, the world's ten thousand or so species can be divided into two camps when it comes to child-rearing: those whose chicks are naked, helpless and require constant care and attention (much like our own offspring); and those whose young are able to feed, walk or swim, and generally keep up with their parents soon after hatching.

The two categories have two different sets of words to describe them, either of which mean more or less the same thing. The naked chicks are known as 'altricial' or 'nidiculous' (from the Latin meaning 'in the nest', whereas the more advanced, self-reliant type is known as 'precocial' (similar to precocious) or 'nidifugous' – meaning 'outside the nest'.

Pros and cons

Birds in the former category include all songbirds and many 'near passerines' such as woodpeckers or kingfishers. Those in the second category include all the offspring of ducks, geese and swans (ducklings, goslings and cygnets), waders, and gamebirds such as pheasant, grouse and partridge.

The precocial strategy does on the surface at least appear more sensible. Surely it is better to give birth to young that have a degree of independence as soon as they emerge from the egg? And it is certainly true that birds living in certain habitats, especially ones that nest on the ground where the chicks could easily be discovered by predators such as small mammals or foxes, tend to be precocial.

But the altricial strategy must work too, or it could not have evolved in so many of the world's species. For although the young are helpless for several weeks after birth, the advantage is that they are confined to the nest, where with luck and judgement the parents can keep them safely hidden away from prying eyes. By the time they do finally fledge and leave the nest, they are often able to look after themselves, and have a reasonable chance of survival.

The downside of having up to a dozen young in a cramped nest is that it can soon become very unhygienic and a breeding ground for parasites. One way in which songbirds in particular have evolved to deal with this is that the youngsters' droppings come ready wrapped in a light, transparent membrane – known as the 'faecal sac' – which keeps the contents from soiling the nest and its occupants.

Half and half

Some birds follow a halfway-house approach. For example, young gulls are 'semi-precocial': born with downy covering and open eyes, but remaining in the nest and being fed by one or both parents for several weeks afterwards.

Even when the chicks have fully fledged (i.e. gained their first 'proper' covering of feathers) and left the nest, many are still heavily reliant on their parents. Many songbirds such as Long-tailed Tits travel in family groups, the young chicks often begging their parents for morsels of food by fluttering their wings in a ritualized begging gesture. Others, including larger seabirds such as albatrosses, may be fed by their parents for many months – even a year or more – after hatching!

ABOVE: Waders such as this Ringed Plover use the 'broken-wing strategy' to lure predators away from their young.

OPPOSITE: A male Corn Bunting may have as many as seven different females on the go at the same time!

Protect and survive

Parent birds have all kinds of clever strategies to keep their young safe. Hiding the nest away is just one option, but what happens once the chicks start to explore the wider world, when they are at their most vulnerable to predators?

One ploy is camouflage: many precocial species have offspring whose downy covering is a blotchy grey or brown – perfect for blending in with pebbles on a beach. But the parent knows that a sharp-eyed predator may eventually be able to pick out a chick, especially if it moves. So several species of wader, including the Ringed Plover, have adopted the 'broken-wing strategy': if a predator or potential threat (including ourselves) approaches close to the chicks, the parent bird stretches out its wing as if it is broken and walks as close as it dares to the predator, often uttering a plaintive call as it does so.

Distracted by this seemingly easy meal, the predator pursues the 'injured' bird and is thus led away from the chicks, whereupon the parent too flies off to safety.

Unusual breeding behaviour

We tend to imagine birds as pairing for life, being monogamous and entirely faithful; and indeed many of them, such as Mute Swans, basically do and are. But the relationships that go on in the world of breeding birds would be shocking were it not for the fact that they are not games but the serious business of maximizing their reproductive success.

DOING IT DIFFERENTLY

Many species of bird are essentially polygamous, with either a male mating with more than one female (known as polygyny) or a female mating with more than one male (less common, and known as polyandry). Many songbirds are basically polygynous, with some males having six or seven different mates. However, even when a female is basically faithful to a single male she may still occasionally go for 'a bit on the side' – mating with an unpaired male in order to maximize her chances of laying eggs of the highest eggs.

The reason that all females are not as brazen as their mates is that it is to the female's advantage to make her mate feel secure, so that he will stay around and do his share of the egg-brooding duties (or at least feed her while she is incubating). Afterwards she also needs him to help look after and feed the chicks; though not all males do so.

Some birds are so desperate to mate with anything even vaguely suitable that they end up with a partner from a different species altogether. Usually, however, they choose one from at least the same family, typically a close relative. This leads to hybrid young being born, and is particularly common among wildfowl, especially geese and ducks. However, these young are generally infertile and so the line soon dies out, although sometimes not before they have caused great confusion among birdwatchers, baffled by a strange-looking bird.

CUCKOO IN THE NEST

Of all Britain's 220 or so species of regular breeding bird, the prize for the most bizarre behaviour surely has to go to the Common Cuckoo. It is the only British bird whose entire breeding strategy is based on cheating: on laying its eggs in other birds' nests, and getting them to do all the hard work of raising the young.

The technical term for this behaviour is 'brood parasitism', and although the Cuckoo is the only British species which habitually practices it, this strategy is followed by about 100 of the world's species of bird, including many other cuckoos, the cowbirds of North and South America, and other tropical species.

Parasitism pays

It may seem a odd way of raising a family – but it does work. After all, because the female Cuckoo does not have to put any energy into rearing her offspring – nor even find a nest big enough to hold all her eggs – she may lay as many as forty eggs in a single year.

Being parasites, Cuckoos need hosts. In the case of the Cuckoo in Britain these mainly come from four common species: the Meadow Pipit, Reed Warbler, Dunnock and Robin. A female Cuckoo will generally only lay her eggs in the nest of a single host species, and therefore is able to match the colour and pattern very well in most cases (though not generally to that of the Dunnock, which may be a relatively recent host species).

BELOW: The Reed Warbler is one of the main host species in Britain for the Common Cuckoo.

OPPOSITE: Male Ruffs posture to each other in a communal display ground known as a 'lek'.

Hawk look-alike

Birdwatchers sometimes mistake a low-flying Cuckoo for a bird of prey such as a Sparrowhawk. Some scientists believe this is no accident, and that the Cuckoo has evolved its streamlined shape in order to flush small birds off their nests. In the confusion, the Cuckoo can perch briefly over the empty nest, remove a single egg, deposit her own egg, and fly away. The unsuspecting host species returns, sees the usual number of eggs, and continues to incubate as if nothing has happened.

Cuckoos only take a short while to hatch, usually emerging before the host's chicks. In this way the baby Cuckoo can actually eject the other eggs (or sometimes helpless chicks) from the nest and take the lion's share of the food for itself. It will need it: by the time it fledges it must grow to several times the size of its host's offspring, filling the nest like some overweight teenager constantly demanding food. Once fledged, it leaves the nest and its exhausted foster parents, and a few weeks later undertakes the extraordinary journey south to Africa – without any help from its true parents, which it never sees.

Playing the field – lekking

A handful of bird species have a really bizarre form of courtship display, known as 'lekking'. Leks involve unmated males competing with each other in front of watching females. The males return day after day to the same arena, performing ritual dances,and defending tiny patches of land to gain some advantage in the race to reproduce. From time to time, a female will allow a male to mate with her. Afterwards he has nothing to do with the nest-building, incubation or raising the young – he simply returns to show off again with his ever more elaborate displays.

Leks evolved in places where there is widespread and readily available food, so defending a territory has no advantage. If this were not the case, lekking males would soon be superseded by any male able to corner a limited food supply.

One of the most amazing lek displays is by the White-bearded Manakin, found in Central and South America. This tiny black-and-white bird dances, clicking its wings loudly like a miniature firecracker, around a tiny clearing in the understorey of the rainforest. Great Snipe, in Poland, take a different approach: they leap high into the air from dense ground vegetation. In Britain, Black Grouse dance around each other like prizefighters, lunging occasionally but rarely causing injury.

WHERE BIRDS LIVE

The distributions, populations and ranges of birds are complex subjects which can only be touched on in a work like this. Nevertheless they are important to include; it is only by knowing the background to where a particular species lives now, and any changes in its range or the size of its population, that we can fully understand that species.

This chapter begins by examining bird populations and how we try to estimate them on a national and global scale. It then takes a closer look at bird distribution – essentially the reasons why birds choose to live in certain areas, why some regions have more bird diversity than others, and why some species are widespread while others are localized. Changes in range are then examined, including colonizations, local and global extinctions and contractions in range. The final section covers both endangered species and those that have already become extinct.

RIGHT: Penguins are not restricted to the Antarctic. The largest King Penguin colonies are on the Falkland Islands and South Georgia.

Bird populations

Imagine looking at a flock of birds, and trying to count them... ten... a hundred... a thousand – often you can only ever 'best guess' the number of individuals involved. Now imagine doing the same process for Britain's entire bird population, then Europe's, then the whole world's. You cannot possibly see all the birds yourself, or visit all the places where they are found, so you must rely on estimates made by others, at different places, at different times, and using very different methods.

NUMBERS OF BIRDS

Is it any wonder that estimates of the total number of birds in the world vary so widely, from tens of billions through to hundreds of billions of individuals? Indeed, so few people have even attempted to answer the question that the last reliable estimate we have is more than half a century old! In 1951, British ornithologist James Fisher calculated that the total world population of birds was in the order of 100 billion – at the time, about thirty times the number of human beings.

Since then, no one has come up with a better estimate, but we can be pretty sure that as the world's human population has risen to well over six billion, the number of birds has almost certainly gone down, due to changes in land use, habitat loss and persecution. We can only guess that if Fisher's original estimate was correct, then perhaps between fifty and eighty billion individual birds survive today.

Familiar figures

What about estimates for places we know a bit more about, such as Britain and North America? Again, we can only rely on best estimates: ornithologists such as Jim Flegg and Max Nicholson have estimated a total British population of between 120 and 134 million individuals – between one in three hundred and one in eight hundred of the world's birds. In North America, James Fisher and Roger Tory Peterson once guessed that the US population of birds might number as many as 20 billion – rather a lot, if we compare it to the global figure of 100 billion. About five billion seems a more likely number.

We do know that numbers of birds (both species and individuals) tend to increase the nearer you get to the equator, with the tropical rainforests and savannahs having the greatest numbers. Polar regions have lots of birds too, but these tend to concentrate in colonies such as the great seabird breeding islands of the Arctic and sub-Antarctic, with very large numbers (sometimes literally millions) of a few species.

BELOW: Lesser Flamingos congregate in vast numbers on the salt lakes of Africa.

All at sea

This has led some writers (including Fisher) to maintain that the world's commonest bird was likely to be a seabird – Wilson's Storm-petrel being the likeliest contender. However, more recent studies have disproved this theory, and though this diminutive storm-petrel is almost certainly the world's commonest seabird, it is far outnumbered by the vast flocks of some landbirds such as the tiny Red-billed Quelea, now known to be the world's commonest bird (see page 33).

Other very common – and far more widespread – species include the House Sparrow (introduced by man to many parts of the world), European Starling, and Cattle Egret – all supremely adaptable species.

But the real surprise is that the commonest bird that ever lived can no longer be seen anywhere in the world, having become extinct early in the 20th century. The Passenger Pigeon was confined to North America, where flocks are said to have darkened the skies on their flights overhead. The total population has been estimated as anything between three and ten billion individuals – yet during the course of the 19th century the species was driven to extinction in the wild. The last surviving specimen finally gave up the ghost in 1914, in Cincinnati Zoo.

THE UK'S COMMONEST BIRDS

According to the British Trust for Ornithology, the top five British birds in terms of total breeding population are, in ascending order of abundance: at number five, House Sparrow (3.6 million breeding pairs); at number four, Robin (4.2 million pairs); at number three, Blackbird (4.4 million pairs); at number two, Chaffinch (5.4 million pairs); and in the number one spot, the tiny and often overlooked Wren (way out ahead with 7.1 million pairs – plus another 2.8 million pairs in Ireland).

This list does change over time. In James Fisher's *Watching Birds*, published in 1941, he estimated as follows: Chaffinch and Blackbird (1st equal with about ten million individuals each); Starling and Robin (3rd equal with about seven million each); followed by House Sparrow, Dunnock, Song Thrush and Meadow Pipit with about three million each.

LEFT: Once the world's most abundant bird, the Passenger Pigeon of North America became extinct in the early 20th century.

ABOVE: The diminutive Wren is now Britain's commonest bird species, with numbers estimated at more than seven million pairs.

Drop outs

Song Thrush has long since dropped out of the top ten, due to a rapid decline in numbers; while Wren, which Fisher placed at about tenth, is now the undisputed number one! Other species which Fisher had in his 'top 30' included Yellowhammer, Linnet, Lapwing and Grey Partridge, all of which have declined dramatically due to the effects of agricultural modernisation in the decades following the Second World War.

All this goes to show that just as it is difficult to make any sensible estimate of bird populations, so that estimate is out of date almost as soon as it is made. Nevertheless the BTO and its volunteer membership have done a great job in at least allowing us to make sensible estimates of bird populations, from which we can monitor declines.

And a final question: why has the humble little Wren gone from number ten in the list in 1940 to number one in the early 21st century? The answer is simple: a long run of mild winters, going back at least to the mid-1980s and arguably even earlier, has reduced first-year mortality in Wrens considerably. In addition, Wrens are able to produce more than one brood, and an earlier start to the breeding season (again the result of mild winters and earlier springs) may help them in this regard.

Bird distribution

Bird distribution is far from uniform: not only does the number of different species vary considerably from place to place, region to region and continent to continent, but for each species the pattern of how they are distributed varies too. Some birds are only found in very small areas while others cover a large part of the globe.

COSMOPOLITAN SPECIES

Some birds are found right across the world in suitable habitats. Known as 'cosmopolitan', these species may have achieved their global supremacy in different ways. So a seabird such as the Arctic Tern covers the globe (or at least many of the marine and coastal parts of it) via its epic migratory journeys from the Arctic to the Antarctic and back. Arctic Terns have been seen in (or off) every one of the world's seven continents, though they are far from widespread everywhere, being absent from any landlocked region.

Globetrotters

Another globetrotting species is the Cattle Egret, which like the Arctic Tern has also been seen in all seven continents, albeit only as a very rare vagrant to Antarctica. The Cattle Egret is also the world's champion colonist: now breeding in six continents thanks to its amazing ability to expand its range by taking advantage of domestic animals in order to find invertebrate food.

Not all that long ago, Cattle Egrets were confined to the Old World continents of Europe, Asia and Africa, but during the past century or so they managed to cross the seas to Australasia, and even more extraordinarily get blown by a tropical storm across the Atlantic to South America, where they settled, thrived, and expanded their range northwards, finally reaching North America by the 1940s. Today they can be found in most warm temperate and tropical areas of the world.

Other birds with a widespread (though often fragmented) global distribution include the Osprey, found in the Americas as well as Europe, Asia and Africa; the Barn Owl, whose wide distribution is reflected in a large number of distinct races; and the Peregrine.

Helping hand

Each of the species already mentioned colonized the places they live by natural processes, whereas two other fairly ubiquitous species, the House Sparrow and European Starling, did so via human agency. House Sparrows have spread over much of the civilized world by hitching a ride on ships or being taken as pets and then released. Starlings colonized North America by a rather less orthodox route: a number of them were released by an eccentric 19th-century naturalist in New York's Central Park, in an attempt to introduce every bird mentioned in Shakespeare to the New World. His project failed, apart

OPPOSITE: The Osprey is one of the most widely distributed bird speces. Occurring in all continents apart from Antarctica, it is flexible in terms of habitat so long as a supply of fish is on hand.

from one species – the Starling is now one of North America's most widespread, and least welcome, birds!

RELICT SPECIES

The term 'relict' species simply means one whose range is now either patchy or discontinuous, with isolated 'islands' of population occurring in widely separated places across an extensive area. Scientists can infer that when this is the case, the species once covered a much larger range, but over time has disappeared from most parts of it.

Ice Age remnant

One example in Europe is a small finch called the Twite. In Britain the Twite is a northern bird, found on the high moors and remote islands of northern England and Scotland, where it replaces its commoner and more widespread relative the Linnet (indeed the Twite has been given the folk name of Mountain Linnet). Twite also breed in parts of Scandinavia, but are not found as a breeding bird in the rest of Europe. That might not seem anything unusual, but for the fact that from Turkey and the Caucasus mountains the Twite's range extends eastwards to Mongolia, western China and the Himalayas.

So why are there two quite separate populations of this species? It is thought that the Twite originally evolved in western Asia, then spread eastwards and westwards. However, when the Ice Ages blanketed much of continental Europe in snow and ice, the northern and western population became isolated. Even when the ice retreated, the bird remained where it was, rather than recolonising the area in between. Hence today's 'relict' population of Twite in north-west Europe.

Cut off in their prime

Other birds with relict populations include Caspian and Gull-billed Terns, both of which have fairly widely spread populations, but with many areas where they are not found. For example, in North America the Caspian Tern breeds on Great Slave Lake in Canada's North-western Territory; on the Great Lakes; in Labrador in eastern Canada; and in coastal areas of south-eastern states such as Florida. We can infer from this that the bird must have once been distributed across a much larger area, with a more continuous range.

The Shore Lark (more appropriately known as Horned Lark, because of its tiny black horns) is the only lark species found extensively in both the Old and New Worlds. But it also has a number of populations cut off from its main range; for example, in Morocco and the Andes.

OPPOSITE: *The Shore (or Horned) Lark is a very widespread species, although its range is highly fragmented.*

ABOVE: *The tropical rainforests of South America, Africa and Asia support a vast proportion of the world's bird species.*

The tropics and the poles

Birds conform to the rule which says that broadly speaking, the closer you get to the tropical and equatorial regions of the world, the greater the diversity of species; while as you approach either of the two poles, the species diversity drops dramatically. So Colombia, near the equator, has a birdlist of at least 1700 different species (and increasing all the time as some species are 'split' from close relatives); while the number of breeding species on the vast Antarctic continent is less than a dozen, with another few dozen occurring as visitors or vagrants.

The reason for this small number of species at the poles is that the habitat is very uniform, with a resulting low level of biodiversity. However, what the poles lack in variety, they make up for in numbers, with vast colonies of penguins, tubenoses and other seabirds, sometimes numbering over one million individuals.

Equatorial and tropical regions support a very varied avifauna because their habitats are not only very varied, but are often ecologically isolated from one another. This means that the different areas act as 'islands', with evolution progressing at a very rapid rate once a population of a species becomes isolated from another one. Tropical rainforests support a highly diverse range of species and new ones are being discovered every year.

ENDEMICS

An endemic species is defined as one whose range is confined to a particular area: usually a small one such as an island or country, but occasionally the term is used to refer to a much larger range, such as a continent.

Endemics occur most commonly on islands: either real ones (oceanic islands surrounded by sea), or ecological ones (mountaintops, or an isolated area of forest cut off from its neighbouring forested areas by grassland). The key factor in determining how many endemic species there are in a particular area is usually time: islands such as Madagascar or Jamaica, which have been isolated from the mainland for many millions of years, have a much higher proportion of endemics that those such as Britain and Ireland or Trinidad & Tobago, which were joined to their neighbouring continent until relatively recently – in Britain's case only 8,000 years ago.

In Madagascar, around one third of all bird species are endemics, found nowhere else on earth. In New Guinea the proportion is even higher – about half the total species are unique to the island. Australia itself has more than 200 endemic species, reflecting how long the continent has been cut off from other land-masses; however, in the whole of the United States of America (excluding Alaska and Hawaii) there are only about ten, including one of the world's best-known birds, the California Condor. If Hawaii is included, the number of US endemics increases by more than thirty species.

Inaccessible islands

Oceanic islands have some of the most interesting endemic species, including several flightless ones. Of these, perhaps the most peculiar is the tiny Inaccessible Island Flightless Rail, found on the speck of rocky land of the same name, in the middle of the south Atlantic Ocean, over a thousand miles from any continental land-mass.

Other famous island endemics include the birds of the Galapagos Islands, lying in the Pacific Ocean off the coast of Ecuador. It was these birds, notably the set of highly adaptable passerines known as Darwin's finches, which helped Charles Darwin formulate his theory of natural selection. These finches are not only endemic to the islands of the Galapagos; they also display a trait common to island forms known as

Wrens

The Wren (known in North America as the Winter Wren) is one of about 75 members of its family, all of which, (apart from ours) are found in the New World. Wrens as a family probably evolved first in North America, then spread southwards into Central and South America, where many species can be found today.

One pioneering wren species managed to cross the Bering Strait (probably via a land bridge) into eastern Asia, from where it has spread westwards, until today it can be found throughout Europe and Asia, as far west as Ireland.

Throughout its extensive range the Wren evolved into numerous races or sub-species. In Britain alone there are several distinct island races, found on Shetland, the Hebrides, and on two tiny specks of land – Fair Isle (between Orkney and Shetland) and the isolated island group of St Kilda, off the Western Isles.

These island races are larger and darker than their mainland cousins, with louder and more distinctive songs. On St Kilda the Wren nests in the old stone walls and buildings left by the people who once lived there, living up to its scientific name of *Troglodytes troglodytes* – meaning 'cave-dweller'.

'adaptive radiation', in which a single ancestral form evolves very quickly into a number of different species, each exploiting a vacant ecological niche.

Volcanic splits

The Canary Islands show an interesting pattern of endemism. Lying a couple of hundred miles off the coast of north-west Africa (though politically and economically tied to Spain), the Canaries boast at least eight endemic species (possibly more, depending as always on which taxonomic principles you choose to follow). These include the famous Canary (later bred as a cagebird), the Blue Chaffinch, and two species of laurel pigeon. All these are found on the island of Tenerife.

OPPOSITE: The Crested Coua, a member of the cuckoo family, is confined to the island of Madagascar.
RIGHT: Wrens are one of the few bird families to have colonized the Old World from their original home in the Americas.

Range changes

Bird distributions fluctuate over time – some contract while others expand. Natural range extensions occur when successful species spread and colonize new areas, but sometimes they have a helping hand. Not all birds that live in a particular place are native there: either by accident or design, human beings have frequently introduced birds that belong in one part of the world to another. Humans have also been responsible for massive contractions of birds' ranges and may yet rewrite the bird maps further through the effects of climate change.

INTRODUCTIONS TO BRITAIN AND NORTH AMERICA

Of course, the process of introducing birds (and indeed other creatures) to Britain began long before Columbus sailed the ocean blue. The first documented introduction to Britain was the Pheasant, now one of our commonest and most widespread birds. Originally from Asia, Pheasants were brought through Europe to Britain by the Romans, who presumably enjoyed a good meal as much as anyone.

Many centuries followed without other species being added to the British avifauna; indeed, it was not until we began to explore other parts of the world and extend the British Empire that the first were brought back, usually as ornamental additions to a stately home and its gardens. One of the first of these species was the now ubiquitous Canada Goose, which as its name suggests is normally found in North America. Canada Geese were first introduced here in the 17th century by rich noblemen as a decorative addition to their lakes. By the late 20th century the birds had reached pest proportions in some locations, dominating other waterfowl and ruining the grass of parks with their constant defecating.

Ruddy ducks!

Yet another North American import, the Ruddy Duck, has arguably caused even more trouble than the better-known Canada Goose. First brought to the Wildfowl Trust headquarters at Slimbridge, Gloucestershire, by the great conservationist Peter Scott, a few pairs escaped from their pens and began breeding locally. At first, no-one imagined that they would succeed, but soon they had spread north and east to the Midlands, where they found a suitable home on the various gravel pits and reservoirs of the region. For most of the second half of the 20th century they were just another curious addition to the British scene, but then we began to hear reports from Spain that Ruddy Ducks were causing havoc with a rare native species. Apparently they were breeding with the endangered White-headed Duck, and were so dominant that the latter species was severely threatened. Since then the controversy has ranged back and forth, with some conservationists advocating a total cull of the birds and others adopting a 'live-and-let-live' approach.

Not all ornamental wildfowl have been quite so controversial. Mandarin Ducks, originally from China and Japan, were introduced in the 19th and early 20th centuries, mainly in the Home Counties around London. They have not expanded their range quite so dramatically as the Canada Goose and Ruddy Duck, but have managed to not only survive but thrive; as has another exotic species, the Egyptian Goose. Meanwhile, a range of other species, including Little Owl and Red-legged Partridge (from southern Europe) and Rose-ringed Parakeet (from northern India) have also managed to become well established.

American imports

Meanwhile, the British are responsible for several introductions to North America: some relatively benign, such as the Grey Partridge; others definitely less so, such as the House Sparrow and Starling.

But the prize for the most bizarre introduction to the North American continent

OPPOSITE: Now a common British bird, the Pheasant was probably brought here by the Romans as a source of food.
BELOW: The European Starling has been one of the most successful introduced bird species in North America.

birds were introduced to the East Humboldt mountains of the state of Nevada from the early 1960s onwards, in order to be shot for sport. Amazingly a few have survived, and are keenly sought by American listers eager to get them on their life list!

must surely go to the Himalayan Snowcock, a bird whose natural range, as its name suggests, is the rocky slopes of the Himalayan mountain range. These

GLOBAL INTRODUCTIONS

While most of the bird species introduced by man to Britain and North America have been relatively benign (or at least not disastrous) to the local ecosystems, the same cannot be said for other introduced species around the world. In

Parakeets

The population of Rose-ringed Parakeets around the London suburbs is now a well-known phenomenon, thanks to a combination of their bright plumage, noisy screeching call and habit of feeding in gardens, all of which have guaranteed this bird plenty of exposure in the media. But a mystery remains: how did these birds, originally from northern India, get here in the first place?

The first parakeets seen in the wild (apart from the odd escaped cagebird) were discovered in Dartford, Kent, in 1969. Reports followed soon afterwards of flocks at large in various suburbs of west London, including Wraysbury and Shepperton. The presence of parakeets at the latter locality led to the myth that they had been released at Shepperton Film Studios during the filming of *The African Queen*, the blockbuster movie starring Humphrey Bogart and Katharine Hepburn. It sounds plausible except that the film was made in 1951 and the chances of a flock of bright green parakeets remaining undiscovered for 20 years is extremely unlikely.

Wraysbury and Shepperton are both near Heathrow Airport, hence the theory that the birds escaped from a delivery there. Surely it is most likely that birds escaped or were deliberately released in several localities by aviary or pet-shop owners, they found the environment to their liking, and not only survived but thrived, becoming a true British bird.

OPPOSITE: Rose-ringed Parakeets are now a regular sight in parts of Britain, having arrived originally as cagebirds.

ABOVE: The humble House Sparrow may be declining in Britain, but it is doing remarkably well globally.

some places – especially remote offshore islands and archipelagos – the introduction of new bird species has been nothing short of disastrous for the local wildlife, and sometimes the people too.

House Sparrows and Starlings cause massive crop damage in places where they have been introduced, such as North America and parts of Asia and Australasia. This is ironic, especially as the sparrows were originally introduced by American farmers to control insect pests! House Sparrows are the most successful (from their point of view at least!) introduced bird, their non-natural range now comprising more than half their original one; indeed their range covers more than one-quarter of the world's land surface.

Serious competition

Parakeets of various species – which as popular cagebirds tend to turn up in ports and cities everywhere – can have a serious effect on other hole-nesting birds, as their aggressive habits mean that they tend to dominate in competition for nest spaces. They are especially common in the Miami area of Florida (where dozens of species have been accidentally introduced by cagebird lovers).

But the problems faced in Britain and America are insignificant compared to the disastrous ecological effects of introductions to two of the world's most remote areas: Hawaii and New Zealand. In Hawaii, more than 160 non-native species have been introduced, causing havoc among the delicately balanced ecosystem, though only about 30 still survive. Well over 100 species (mainly from Europe and Australia) have been introduced to New Zealand, and visiting birders often remark on the abundance of Song Thrushes and Goldfinches there.

Easy exploitation

Many of these non-native birds are able to thrive, partly because of the lack of mammalian predators, but mainly because they are already well adapted to exploiting agricultural practices and living alongside human beings – something the native birds have not had time to get used to. As a result, many native New Zealand species are in severe decline, often confined to remote offshore islands which non-native birds (and other alien invaders such as rats and cats) have not managed to reach.

Meanwhile Australia has also had its fair share of non-native species, including the Common Myna and Indian Spotted Dove, both from Asia.

BELOW: Now a familiar garden bird, the Collared Dove only arrived in Britain half a century ago.

Overall, introduced birds have not proved quite so destructive as creatures from other animal groups; notorious examples are the Grey Squirrel (from North America) to Britain, the Nile Perch in the Central African Lakes, and the Cane Toad in Australia. Nevertheless, authorities are now much more careful to try to prevent the accidental or deliberate introduction of any species from one region of the world into another, to try to minimise problems in the future.

COLONIZATIONS – NEW KIDS ON THE BLOCK

Some birds do manage to extend their ranges on their own – sometimes quite dramatically. During the last century, the two best examples are those of the Collared Dove and Cattle Egret, both of which colonized new areas (in the case of the Cattle Egret, new continents) by their own, natural means – though in both cases they took advantage of new ecological niches provided accidentally by human influence.

The story of the Collared Dove is a well-told one, but is so amazing it is worth repetition. At the start of the 20th century it was basically an Asian species, breeding no further west than the Ural Mountains of south-west Asia. By the 1930s it was breeding in Turkey, after which it began to spread very rapidly north- and westwards across the mainland of Europe. It is thought that the doves took advantage of an increase in the amount of grain available on farms, though whether or not this in turn resulted from a change in agricultural practices it is hard to say.

British breeder

Whatever the reason, the doves were well able to take advantage of a new ecological

ABOVE: The Cattle Egret is a likely contender for the next newcomer to the British breeding avifauna.

niche with few competitors. In Britain, the first birds were sighted in Lincolnshire and Norfolk in the early 1950s, with a pair breeding on the North Norfolk coast soon afterwards. By the 1960s they had spread throughout much of southern Britain, and less then fifty years later were firmly established as one of our most widespread, common and familiar birds. Today the population has declined a little (perhaps due to modern farming methods which leave less spilt grain), and the bird has adapted to living in suburban and rural gardens, where it thrives.

Collared Doves have even established a toehold across the Atlantic in North America, with birds breeding in various eastern states. However, we cannot be sure if these are descended from wild birds which had flown across the Atlantic under their own steam, or from a free-flying population introduced to Bermuda.

Egret extension

The Cattle Egret is also now a common and familiar sight in the Americas, being found over much of the North and South American continents. In this case, we do know for sure that the original colonists did get there under their own steam. Sometime in the early part of the 20th century a severe storm carried a flock of egrets across the Atlantic from east to west. The birds, originally from western Africa, landed in Guiana on the north-east coast of South America. Here, a fledgling cattle-farming industry was getting underway and the egrets took full advantage of this, feeding on the invertebrates churned up by the cattle's hooves or attracted by their dung.

Soon, Cattle Egrets were being seen in Florida, and by the 1950s were a regular sight there. Afterwards they spread rapidly throughout the warmer regions of North America and are now common and widespread.

DECLINES AND CONTRACTIONS IN RANGE

Meanwhile, as some successful species extend and expand their ranges across the globe, others are in steep decline. Many of these are specialized species whose range was never very large; but others were once common and widespread species which simply found it too hard to adapt to the massive changes brought about by human beings and our insatiable need to transform the world and its habitats for our own short-term use and benefit.

In Europe, and to some extent in North America, the biggest and most severe declines in numbers and contractions in range have been seen among farmland birds. This is a direct result of modern farming methods, introduced after the Second World War when maximising production and especially crop yield was given complete priority over any other factors such as the welfare of our wildlife.

Poisonous pesticides

The first – and still the biggest – catastrophe came in the 1950s and early 1960s, when Rachel Carson's polemical book *Silent Spring* brought the pesticides scandal to our attention. Since the end of the Second World War, scientists had developed powerful pesticides known as organo-chlorines, of which the best known (or at least the most notorious) was called DDT. This was hailed as a miracle breakthrough by farmers, as it was highly effective in killing insect pests that destroyed their crops and reduced their precious yield.

But DDT had two major problems: first, by killing insects indiscriminately it reduced both biodiversity and the amount of insect food required by many birds. This in turn made it harder for them to raise young, resulting in fairly rapid population declines and contractions in range. Second, and even more seriously, DDT concentrated and persisted in the food chain, so that the higher creatures in the chain (the large predators such as raptors) accumulated very high levels of the poison in their bodies. In the case of the Peregrine Falcon, this had a disastrous effect on the thickness of their eggshells, with the result that eggs with thinner shells failed to hatch. Within a few years the Peregrine population on both sides of the Atlantic had plummeted, and the bird faced localized extinctions.

Fortunately the problem was spotted just in time, and after campaigns by ornithologists and the general public DDT was withdrawn from use, though it is still licensed to many areas of the developing world where it continues to wreak havoc.

Farmland declines

Pesticides are not the only factor affecting farmland birds. During the same period the farming landscape has changed from small, mixed farms with plenty of hedgerows and weedy fields to vast monocultural deserts of wheat or barley – totally unsuitable for birds, flowers and insects. In Britain, this resulted in rapid declines for many familiar farmland birds, especially the Skylark, Lapwing, Linnet, Yellowhammer, Corn Bunting, Tree Sparrow and Grey Partridge. Fortunately, the trend now appears to have been reversed, with governments encouraging farmers to consider wildlife and even encouraging them to do so by means of grants. It is too early to say whether this change in direction will come in time to save some of our best-known and best-loved species of bird.

CLIMATE CHANGE AND RANGE EXTENSIONS

Over the coming few decades, the greatest changes to the ranges of the world's birds are likely to be as a result of climate change. Though the long-term impacts of global warming on our weather and climate are highly unpredictable, the current trend – in Britain and Europe at least – towards milder winters and hotter summers is likely to see some major shifts in the distribution of many of our bird species.

Because Britain is on the edge of a vast landmass, and in a temperate zone, many of our breeding birds are either on the northern or southern edges of their range. If global warming results in significant shifts in temperature we are thus likely to see some dramatic changes in the pattern of distribution for such species, with colonisations and extinctions in our breeding avifauna as a result.

OPPOSITE: Grey Partridges have undergone a population and range decline in Britain due to modern farming methods, but may now be recovering.
RIGHT: The Tree Sparrow is another farmland species which has suffered a major collapse in numbers.

Fall and rise of the Corncrake

In the poet John Clare's day, in the early 19th century, the Corncrake was one of Britain's most widespread birds, found in lowland farmland from southern England to the Shetland Islands. Hardly ever seen, its repetitive, two-note call (which gives the bird its scientific name *Crex crex*) was a familiar sound to rural people everywhere.

The Corncrake is a rather odd member of the rail family as, unlike its relatives, it prefers living away from water (hence its alternative name of Land Rail). It has simple requirements, nesting in crops such as hay, or in wet meadows with plenty of insect food for its young. As farming methods changed, mechanised cutting with combine harvesters was introduced, and the Corncrake was doomed.

Machines, drainage and intensive crop production meant that Corncrakes did not stand a chance. Their range declined rapidly, so that by the end of the 20th century they were confined as a British breeding bird to a few remote islands off the Scottish coast where traditional farming methods were still being practised. A similar decline has occurred over much of Europe where the Corncrake is now classified as threatened.

But now sympathetic farming methods, along with reintroduction schemes in the East Anglian fens, mean it may not be long before the call of the Corncrake is heard once again in its former haunts.

Good for some

Global warming might be good news for some of our breeding species whose range just extends from Europe into Britain, such as Cirl Bunting, Dartford Warbler and Nightingale. As resident species, the first two already benefit from milder winters, with mortality rates much lower than during hard, frosty ones when food is harder to find. So since 1963, when the 'Big Freeze' killed off all but a dozen pairs of Dartford Warblers, the species has

LEFT Formerly widespread, the Corncrake vanished from most of Britain in the face of modern agriculture. The species is now making something of a comeback.
OPPOSITE: Black Kites – the world's most successful raptor – are likely to colonize southern England in the fairly near future.

undergone a major population growth and can now be found on lowland heath throughout southern Britain. If the trend continues the species may well spread northwards into the Midlands very soon, provided it can find enough suitable habitat on which to breed.

Cirl Buntings have declined in the past few decades, like so many seed-eating species, as a result of modern farming methods. Now, however, thanks to careful conservation work, the species is increasing again and could soon return to its former haunts.

New colonists

In the past thirty years or so, three species have colonized Britain from mainland Europe: Cetti's Warbler, Mediterranean Gull and Little Egret. All three were once found mainly in southern Europe, but have extended their ranges gradually northwards, helped by a trend towards warmer climates.

In the next few years they may be joined by other new colonists. Spoonbills have already nested several times, and may soon become a permanent addition to our avifauna (or in this case, a readdition, as they once bred across the East Anglian fens before these were drained). Hoopoes are an occasional breeder, as are European Bee-eaters, which nested successfully for the first time for almost fifty years in a Durham quarry in 2002. Another pair attempted to nest on the River Wye near Hereford in 2005, but unfortunately their chicks were eaten by a predator.

Waiting in the wings

There are plenty of other species which are relatively common and widespread on the other side of the Channel, yet which cannot seem to make the leap required to breed permanently in Britain. Serins nested in Devon and Dorset for several years from the 1960s onwards, yet never quite managed to make their colonization permanent. Zitting Cisticolas (formerly known as Fan-tailed Warbler) breed commonly in northern France and Holland, but remain very rare indeed here. Melodious and Great Reed Warblers could also extend their ranges northwards if conditions were right.

But perhaps the likeliest colonists are two of the world's commonest and most adaptable species: Black Kite and Cattle Egret. Both are widespread throughout France, having moved northwards during the 20th century; both are very versatile; and both do turn up as vagrants to southern Britain, usually in spring. So it would not take much for them to settle down and nest here permanently, becoming just the latest additions to the constantly changing British avifauna.

Climate change and range contractions

As we gain species at the southern end of the country, we are likely to lose some in the north. The most famous Arctic breeding bird in Britain, Snowy Owl, was sadly only a temporary colonist, breeding on the Shetland island of Fetlar from 1967 to 1975, with a few hanging on for a decade afterwards. Snowy Owls colonized because of a brief period of climatic cooling in northern Britain, which brought conditions more like their native Scandinavia. Once things began to warm up, however, conditions were no longer suitable and the birds headed back north.

Fetlar is also home to another species more at home in Iceland and Scandinavia, the Red-necked Phalarope. This tiny wader (known in Shetland as 'pirrie duc' – meaning little duck – because of its habit of swimming) is at the very southern edge of its world range here, and may not survive if global warming takes hold.

But the most serious threat comes to our three classic birds of the high mountain-tops of Scotland: the Ptarmigan, Dotterel and Snow Bunting. All three species breed in Britain at high altitudes, away from competition and most predators, and where suitable food is abundant. Sadly most scientists now predict that as the temperature rises, so this 'arctic-alpine' habitat will disappear, along with three of our rarest and most treasured breeding birds.

BELOW: Climate change may well see the demise of the British breeding population of Snow Bunting.

OPPOSITE: Spix's Macaw is now almost certainly extinct in the wild. A handful of birds remain alive in captivity.

Endangered species

As the world's human population continues to increase and we demand an ever greater share of the earth's natural resources, so the threat to the world's wildlife becomes ever more serious. Birds are in the frontline of this battle between man and nature, living as they do in such a diverse range of habitats and locations. So it is hardly surprising that so many species are now considered endangered, with many possibly even at risk of impending extinction before the end of this century.

THREATENED BIRDS

In the year 2000, the worldwide bird conservation organization BirdLife International published a weighty volume documenting all the world's birds in danger of decline and extinction. *Threatened Birds of the World* covered almost 1200 bird species, placed in different categories depending on the threat the authors considered that they faced.

Almost 700 species are considered 'Vulnerable', with the need to monitor their progress and be aware of any sudden or severe declines. Just over 300 are in the next category, 'Endangered', which means they face a serious risk and need urgent conservation work in order to safeguard their future.

Most worryingly of all, over 180 species were classified as 'Critical', the most serious category and reserved for birds regarded as almost certain to become extinct unless major interventions occur very soon. This category includes some species that are almost certainly already extinct, such as the Eskimo Curlew, a small migratory wader of the Americas that has not been reliably recorded for at least 20 years. Other 'Critical' species include Spix's Macaw, the very last wild specimen of which disappeared from its home in north-eastern Brazil only a few years ago.

Some hope remains

Not all hope is lost, even for the most critically endangered birds. Some species once considered on the brink of extinction have made comebacks, always with the help and support of human intervention.

This is ironic, since their decline was always as a result of human factors such as habitat loss, persecution and pollution. Other species, such as Gurney's Pitta, hang on in the rainforests of Thailand and Myanmar (formerly Burma), with conservationists struggling to preserve the few that remain against a tide of habitat destruction driven by economic pressures.

Almost all species at risk of extinction live in the developing world, with two countries – Brazil and Indonesia – having the most species at risk, each with more than 100 threatened species.

LEFT: A captive breeding programme has brought back the California Condor from the very brink of extinction.
OPPOSITE: Despite major conservation efforts, Kirtland's Warbler is still threatened with extinction.

Back from the brink

In conservation there are some success stories to give us hope as birds once considered doomed to extinction have come back from the brink at the eleventh hour. Probably the best-known examples come from North America, where the continent's largest flying bird (California Condor) and tallest bird (Whooping Crane) have come back from populations numbering only a few individuals.

The statuesque Whooping Crane breeds only in one location in Canada, and winters in Texas, so is highly vulnerable. By 1938 there were only 14 individuals left alive, but conservationists have now increased this to almost 200, with a further 60 or so in an introduced population in Florida.

The situation with California Condor was so critical that all remaining wild birds were taken into captivity, where an intensive breeding programme was begun. This too has been a success, and today free-flying descendants of these captive birds are at large in California and Arizona's Grand Canyon. However, the species has yet to breed again in a wild state.

Other conservation successes include the Mauritius Kestrel, whose wild population fell to just four wild birds in 1974. Thanks to intensive efforts involving captive breeding and scientific study it has reached a peak of between 500 and 800 individuals at three separate sites on the Indian Ocean island.

Kirtland's Warbler

Kirtland's Warbler is a paradox: it is both a success story and a warning for the future. Named after the 19th-century ornithologist Jared Kirtland, this attractive little bird is confined as a breeding species to a single area of jack-pine forest in north and central Michigan, USA. After breeding, the birds migrate south-eastwards, wintering in the Bahamas and Turks and Caicos Islands in the northern Caribbean.

Kirtland's Warbler is, on the face of it, a conservation success. In 1987, the population was reduced to fewer than 200 singing males but now, thanks to careful conservation work, there are more than 1,400. The comeback was the result of understanding the bird's highly specialized habitat requirements: it nests only in stands of jack pine of a particular height – between 6½ and 13 feet (two and four metres) tall, on sandy soil. Regular burning of areas of forest ensures that this optimum height is maintained; a concerted campaign against Brown-headed Cowbirds, which lay their eggs in the warblers' nests, has also helped.

Now the species faces a new problem. Global warming threatens to make the small area of jack-pine habitat unsuitable for breeding, and because Michigan is on the southern end of the Great Lakes, it is unlikely that the habitat can 'shift' northwards. So despite all efforts, Kirtland's Warbler may soon become a casualty of climate change.

Extinction

The Dodo and Great Auk are the best known of more than 80 species of bird that have become extinct in historical times (i.e. after the year 1600). Many of these were confined to oceanic islands and some were flightless. Extinction appears to run in families, with rails, parrots and pigeons accounting for almost half the total of extinct species. Rails were especially vulnerable because so many species lived on tiny islands and were flightless; parrots were affected later on, when so many were captured to supply the cagebird trade.

EXTINCT ICONS

When we think of extinction, the creature that almost inevitably comes first to mind (apart perhaps from the dinosaurs) is that proverbial icon of permanent annihilation, the Dodo. Confined to the island of Mauritius, in the Indian Ocean, the Dodo was a huge flightless member of the pigeon family, related only to the Solitaire of another Indian Ocean island, Réunion. Dodos did not fly because like so many island species, they did not need to; the lack of ground-dwelling predators meant that they could dispense with the power of flight, thereby using less energy.

All was well and good, until the arrival of sailing ships from Europe in the 16th and 17th centuries. These brought sailors hungry for fresh meat after months of dried and salted food; with them came dogs, cats and rats. All of these were adept at hunting a slow, heavy, flightless creature. Sailors ate Dodos by the hundred, until the bird became too scarce to find, after which time it was doomed. The very last Dodo probably died out in the late 17th century, to become forever celebrated as the definitive axiom of extinction – 'as dead as a Dodo'.

Another iconic extinct bird, especially in the northern hemisphere, is the Great Auk. Easily the largest of its family, the Great Auk was also flightless; as with the Dodo, this factor meant it was highly vulnerable to attack by man. Again, sailors were to blame; and when the Great Auk became too rare to be worth hunting for food, its status as a collector's item meant that it was even more sought after by greedy hunters eager to make a fast buck. The very last Great Auks were killed on the island of Eldey, off the southern coast of Iceland, in June 1844.

Rediscovered species

In early 2005, a piece of news was released which astonished birders and ornithologists worldwide. The Ivory-billed Woodpecker – a huge and impressive species considered lost for ever – had apparently been rediscovered in the swamps and backwoods of Arkansas. Last sighted in the USA in the 1940s, it had hung on in the forests of Cuba until the 1980s, but with no reliable sightings for so long it was assumed that the species was finally extinct. Indeed, some authorities are asking for further proof before accepting that it does in fact still survive in Arkansas.

The Ivory-bill would not be the first species considered lost forever to be found again, and nor would it be the last. The rarer a bird gets, the harder it is to see, so judging when it has finally gone extinct is always going to be a matter of informed guesswork. And like all such conjectures, it can sometimes be wrong.

The best-known 'back from the dead' bird is probably the Takahe, a huge rail found in the mountains of New Zealand's South Island, which vanished for fifty years before being found again in 1948.

Some birds disappear for even longer. Jerdon's Courser, a nocturnal species found in northern India, was rediscovered in 1985 after being missing for at least 80 years; and the Invisible Rail was found on Halmahera in Indonesia, also after a long gap without any sightings.

OPPOSITE: The Great Auk was the largest member of its family, yet its inability to fly rendered it particularly vulnerable to human predation. Specimen hunters and egg-collectors hastened its demise, and the last recorded British bird was killed on suspicion of being a witch!

RIGHT: Once thought extinct, the Ivory-billed Woodpecker may have been recently rediscovered in the remote swamps of Arkansas. The number of any remaining birds is likely to be very low and ensuring their survival will be one of the great conservation challenges of the years ahead.

BIRDS AND PEOPLE

It may seem odd to conclude with a chapter which examines the various ways in which birds interact with human beings and have done so over the centuries. But without some perspective on the ways in which the paths and lives of human beings and birds have crossed over the centuries – and continue to do so today – our understanding of birds would be severely limited.

This chapter begins with a brief survey of the historical relationship between birds and people, from the earliest human cultures onwards. The human exploitation of birds for domestication, hunting and pleasure is also examined; this is followed by a section on the place of birds in our culture, including weather lore, the arts, and popular culture. The chapter concludes with sections on conservation and bird protection, and finally on the pastime that so many millions of us enjoy today: birdwatching or birding.

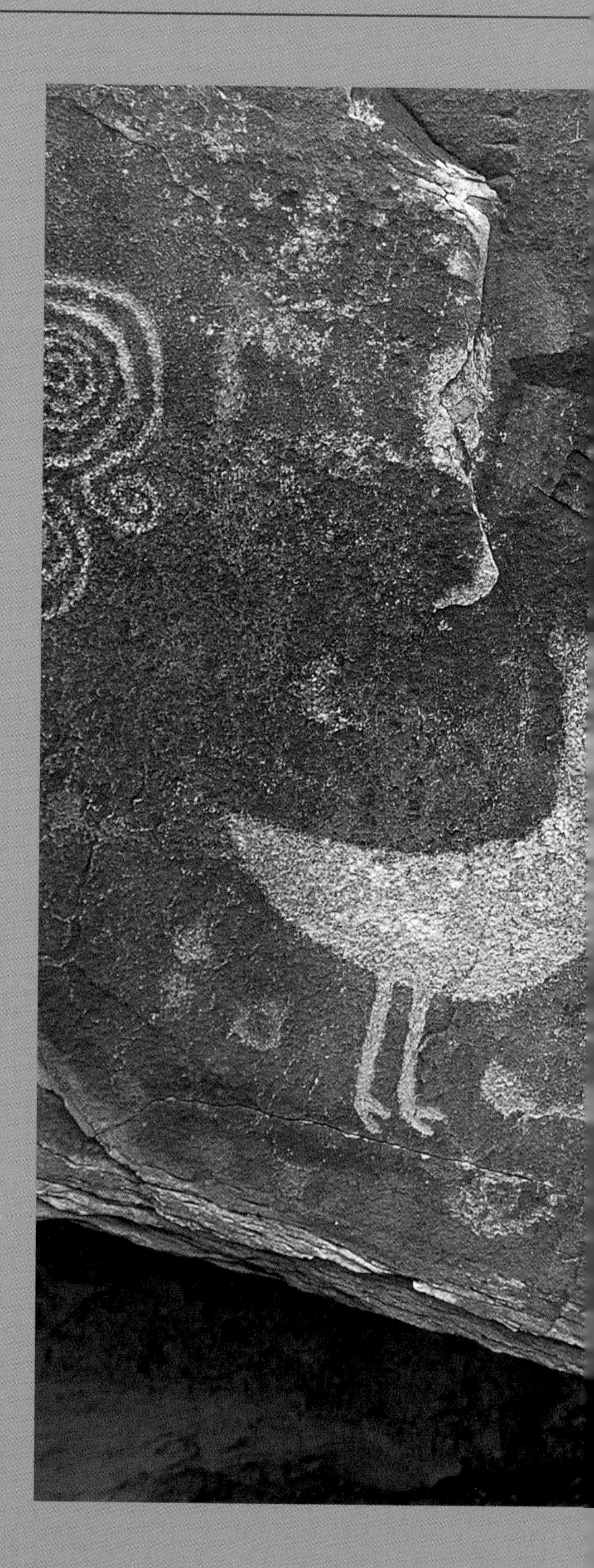

RIGHT: Birds were an integral part of early human life and feature regularly in rock art. These pre-Colombian petroglyphs are located in Arizona, USA.

Birds in history

Very little is known about people's earliest knowledge of, or interest in, birds simply because there are no written records to go by. Only cave paintings indicate their fascination. Early written works by the Greeks, and then in Biblical times, show how man noticed the birdlife around him. Man's association with birds continued through falconry to the naming of different species and the beginnings of ornithology.

ANCIENT HISTORY

Ever since the first prehistoric hunter daubed, carved or scratched an image of a bird onto the surface of the wall of his cave, mankind has been fascinated by birds. He may have been creating this image for superstitious reasons – as an object of worship or to bring him luck in hunting, for example. He may have been doing it for a more practical purpose, to enable himself and his fellow hunters to recognise their prey. Or he may have been driven simply by the impulse to create a work of art which, though primitive in its execution and simple in its form, still resonates when we look at it today.

The very first known image of birds dates back about 18,000 years, to a cave in the French Pyrenees, and depicts a long-legged bird – probably a heron or a crane. Later images from caves in Britain and Spain depict a wider range of species, including waterfowl, bustards and even flamingos. These are, of course, only the ones we know to have survived; other cave artists in different locations might have depicted a very different range of species.

BELOW: 3,000-year-old carvings in the Temple of Amun at Karnak depict a type of wading bird and – on the right – an African Darter, now very rare in Egypt.

Weather lore

We can safely assume that the first primitive farmers took a healthy interest in the comings and goings of birds, either because they might eat their crops or, more positively, because the birds enabled them to foretell the coming weather by observing the pattern of bird behaviour at different seasons of the year. Our current storehouse of weather-based folklore, much of it to do with birds and their behaviour, presumably derives from this period in our history.

We know that the Ancient Egyptians took a keen interest in birds and other wild creatures, though more for religious purposes than any other. Several species are carved on the tombs of the Pharaohs, including those worshipped by the Egyptians as gods, such as the Sacred Ibis. Red-breasted Goose also appears clearly on some carvings, despite the fact that the bird no longer occurs in Egypt.

The Ancient Greeks were celebrated for their great knowledge and spirit of enquiry about the world around them, and birds were not excluded from this. Aristotle, indeed, has a claim to be the first serious student of birds. Living in the 4th century BC, he used his excellent powers of observation to document the behaviour – and especially the migration – of the species living or passing through his region. In doing so he made observations so advanced that they were not repeated for many centuries.

ABOVE: The Raven is the first species of bird to be mentioned in the Bible and appears in the famous story of Noah and his Ark.

Biblical birds

Meanwhile, in another part of the eastern Mediterranean, the authors of the Old Testament proved themselves to be excellent observers of birds. The Holy Land is a crossroads for migrating birds from three continents, so the inhabitants could hardly have failed to notice huge flocks of storks, cranes and raptors passing overhead each spring and autumn.

The very first bird mentioned in the Bible is the Raven, sent out by Noah from the Ark in order to ascertain whether or not he was nearing land. Other well-known Old Testament birds include the dove, stork and eagle: all of which make frequent appearances.

Today, pilgrims to the region can enjoy watching birds at several well-known Biblical sites, including the Dead Sea (home to Fan-tailed Ravens and Little Green Bee-eaters), Masada, where many desert birds are found, and Lake Tiberias (formerly known as the Sea of Galilee), home to a wide range of waterbirds.

The Middle Ages

The study of birds has not always been as prominent in history and culture as it is today. From the ancient world to the 18th century, there was little or no formal study of nature; indeed, the Christian tradition viewed wild creatures as created by God primarily to be exploited by man, so anything written about birds or other aspects of the wild was mainly concerned with hunting.

One way in which birds were used by human beings, and glorified by them at the same time, was through the sport of falconry. Each particular level of royalty or nobility had their own type of bird of prey. The grandest creatures such as eagles were reserved for kings, small falcons such as the Merlin for the ladies of the court, and, as the old rhyme says, 'a Kestrel for a knave', the commonest bird for the lowliest rank of all.

Falconry was especially popular with royalty and one European royal, Emperor Frederick the Second of Germany, took a keen interest in all aspects of bird behaviour, even writing one of the first published books on the subject. But he was in a minority – not until the writings of Gilbert White began to popularise nature study in the late 18th century did people take more than a passing interest in birds.

NAMING THE BIRDS

How do birds get their names? What seems like a simple question in fact has many answers. It depends mainly on the prominence of the bird, when and where it was originally discovered, and in many cases the person who found it in the first place.

Most birds presumably got their name by a process of general agreement, by which I mean that any common bird would have had a wide range of local and folk names, of which a single, popular one would eventually come out on top. It does not take much imagination to see why two common species of warbler were named the Whitethroat and Blackcap – even though the female Blackcap has a chestnut-brown crown!

Sound and action

Likewise, names such as Chiffchaff, Cuckoo and Kittiwake are onomatopoeic – i.e. they imitate the sound made by the bird itself. Many other modern names were originally onomatapoeic too, but with linguistic changes this has been lost. Chough, which was originally pronounced 'chow', is a good example.

Lots of birds are named after what they do: Shoveler (though this was originally applied to the Spoonbill), Treecreeper and woodpecker to name but three. Others were too, though again it takes a little detective work to infer this: Nuthatch means 'nut-hack', while Fulmar means 'foul gull'. Other names reflect a bird's appearance, but this is harder to infer today because the birds were named when Anglo-Saxon was the predominant language. So Wheatear actually means 'white arse', while Redstart means 'red tail'.

Other birds were named after where they live, sometimes inaccurately. Marsh Tit and Willow Tit are easily confused, and indeed the latter was not discovered as a British breeding bird until the late 19th century. Marsh Tit seems a most inappropriate name, and probably resulted from the birds being confused with Willow Tits, which do tend to live in marshy habitats. 'Oak Tit' is surely a more relevant name for the Marsh Tit, while Willow Warbler could be renamed 'Birch Warbler'!

OPPOSITE: A falconer in Kyrgyzstan with his Saker Falcon. BELOW: The Wheatear's white rump led to the name 'white arse'.

Not so common

Some birds were so rare – or at least rarely seen – that they never got a common name. Avocet and phalarope are good examples. Others, now rare, were common enough to get folk names: Red-backed Shrikes were until the mid-20th century fairly widespread across southern England, but are now extinct as a British breeding bird. In the days when they were a familiar sight, their habit of impaling their insect prey on thorn bushes earned them the nickname 'butcher bird'.

Many birds are named after people: usually, though not always, the person who 'discovered' them. About forty species on the British List fall into this category, though most are very rare indeed: only a handful of regular breeders or visitors, including Leach's Petrel and Bewick's Swan, fall onto this category. One of the best known is Montagu's Harrier, named after an English ornithologist of the late 18th and early 19th centuries, who first described this magnificent bird of prey.

Today, bird names are more fixed than they used to be, though from time to time they do change: the bird known today as a Dunnock was until relatively recently called a 'hedge sparrow' – even though it belongs to a quite different family, the accentors. Another problem with many British bird names is that they reflect a time when we were more parochial and insular than today: thus we have had to add prefixes to names such as 'swallow' (now Barn Swallow) and 'cuckoo' (now Eurasian Cuckoo) to avoid confusion with the many other members of these birds' families.

RIGHT: Montagu's Harrier was named some 200 years ago after ornithologist George Montagu. It is Britain's rarest breeding bird of prey.

Human exploitation of birds

Since human beings first settled down and changed from a nomadic, hunter-gatherer lifestyle to one based on agriculture, they have sought to domesticate wild creatures for their own needs and birds are no exception. Birds are also kept in cages as companions, with large numbers now bred in captivity. Wild birds are still caught for food, especially around the Mediterranean where large numbers of migrants are killed.

DOMESTICATION

The first bird to be domesticated – and still the commonest in the world today, was the Red Junglefowl of Asia, which was first domesticated in India in about 3200 BC. The domestic chicken, as it has become, is a highly adaptable form which provides plenty of tasty eggs and a good supply of meat – though in its latest incarnation of 'battery chicken' it has gone from being a palatable luxury to an often inedible, cheap and universally available food, summed up in the 'nuggets' version found in fast-food outlets all over the world.

The global population of chickens now outnumbers even the world's human population: at roughly eight billion vs. just over six billion – though the humans are catching up. Its spread around the world may have happened in both directions: it is known that soon after Christopher Columbus 'discovered' the New World, sailors brought the chicken to Central America. However, some historians believe that the bird may also have arrived from Asia, via the Pacific Ocean, to Ecuador and Peru. Either way, the chicken is now kept – and its eggs and flesh eaten – by societies all over the globe.

Breeding by artificial selection has allowed the original 'chicken' to mutate into many different forms: in the West alone, almost forty breeds are kept for eating, while there are two dozen ornamental breeds as well.

Tasty alternatives

Ducks and geese are also highly popular as domesticated birds, producing a fattier flesh than chickens but equally tasty eggs. Those descended from two wild species, the Mallard and Greylag Goose, are the most commonly kept and widespread; though several other species, including Swan Goose (the domesticated form being 'Chinese Goose'), Muscovy Duck and of course the Turkey (descended from the North American Wild Turkey) are also very popular. The association of Turkeys with Thanksgiving (in North America) and Christmas (in Europe) has kept up the popularity of this huge and impressive bird.

As explorers began to travel the globe, creating trade routes between Europe, Africa, Asia and the Americas, ever more exotic bird species began to enter the domesticated sphere. Guineafowl from Africa are popular as a luxury meat product and can also, like geese, be used as substitutes for guard dogs, as they make a very loud noise when anyone approaches. Another, rather less exotic, domesticated

RIGHT: With roughly 8 billion individuals, the domestic chicken is the world's most common bird.

bird is the pigeon, with all captive birds descended from the wild Rock Dove, a rather shy, cliff- and coast-dwelling species found throughout much of the Old World.

Exotic fare

In recent years commercial use of other wild species has been attempted, not always successfully. Ostrich farming became a craze in late 20th-century Britain, with many people losing money in an ill-advised venture – just how do you boil an Ostrich egg?! Even though the meat is high in protein and very low in fat – so a healthy alternative to beefsteak – people just don't seem to be ready to eat the world's largest bird.

In Britain, some domesticated species have escaped into the wild, breeding freely: Greylag and Barnacle Geese being two examples. In other cases, such as the Rose-ringed Parakeets living wild in London and other parts of the country, the birds originated from the cagebird trade rather than from being kept for food.

BELOW: Named after their place of origin – the Canary Islands – Canaries are one of the world's most popular cagebirds.

THE CAGEBIRD TRADE

For many people, especially those living in cities, their first encounter with a real, living bird will be with a Canary, Budgerigar or one of the other species commonly kept in cages. Ever since the bird artist and ornithologist John Gould first brought back a wild Budgerigar from its native Australia in the middle of the 19th century, people have kept budgies – Britain's commonest cagebird – in cages and aviaries. Today, the industry is a thriving one, though oddly there is very little crossover between people who consider themselves birdwatchers and cagebird enthusiasts, with much mutual suspicion existing between the two groups.

Home-bred varieties

Of course, birds had been kept in cages long before Gould imported his first budgies. Wild birds – especially those that have attractive songs – were frequently captured and kept, often in rather cramped, cruel conditions. Many, especially those that need a regular supply of live insect food such as warblers and Robins, soon died; but those that eat seeds, such as Linnets and Goldfinches, often thrived, and were a popular addition to any Victorian drawing-room. More unusual birds, such as Canaries, were brought back to Europe (from the Canary islands, of course!) as early as the 1500s, and cross-bred with other finch species and each other to produce the innumerable varieties we see today.

Popular parrots

Parrots – mainly from South and Central America, but also from Australia and Asia – have also been a popular cagebird, despite the problems bringing these specialised feeders back from the wild, and having done so, keeping them alive. Sadly, unscrupulous dealers have often taken birds from the wild, which has made heavy inroads into the dwindling populations of several rare species, and may even have contributed to them becoming extinct.

One sad example is Spix's Macaw, which once lived wild in an area of north-eastern Brazil. The wild population dwindled rapidly, creating an even greater demand amongst rich collectors for this beautiful blue parrot. Eventually all but a single male vanished from the wild; until he, too, succumbed (see page 163). Today, a couple of dozen Spix's Macaw's survive, but only in captivity.

ABOVE: Remote communities often depended on birds for their livelihood. Shearwater chicks are still harvested for food in some parts of the world today.

Birders' distaste

Birders tend to dislike cagebird enthusiasts for one simple reason: escaped cagebirds 'muddy the waters' for twitchers, who can never be entirely sure that a rare bird is truly wild and reached Britain in a wild state. In recent years a flood of imports of species from Asia in particular has caused concern for many an avid lister.

Today, however, most cagebird enthusiasts are keen to clean up the image of their hobby and the vast majority of birds held in aviaries in Britain have been captive-bred, rather than taken from the wild.

HARVESTING BIRDS FOR FOOD

Harvesting wild birds for food has a very long pedigree. It is hard to imagine that our prehistoric ancestors, when faced with a thriving colony of edible birds sitting on nests, would not have jumped at the chance to get some easily obtained protein in the form of tasty birds or eggs.

Colonial seabirds were particularly easy to exploit, though their nesting sites, on steep cliffs (Kittiwakes, auks and Fulmars) or down burrows (Puffins and shearwaters), might have given the hunters some cause for concern and considerable difficulties. Nevertheless, the harvesting of wild birds, first simply to get a meal, and later for commercial use, has been going on for many hundreds of years. Manx Shearwater chicks were taken in their thousands and sent to London markets, where they were sold as 'puffins' (which came from the species's scientific name *Puffinus puffinus*). At some point the name became transferred to the bird formerly known as a 'sea parrot', but which we now know and love as the Puffin.

St Kilda

Of all the people of the world, there can be few who harvested and exploited wild birds more effectively than the people of St Kilda. Lying about fifty miles off the Outer Hebrides, on the north-west coast of Scotland, this island group supported several hundred people until the late 19th century, after which diseases and the lure of the outside world gradually reduced the population until the final few were evacuated in 1930.

Despite being surrounded by oceans teeming with life, the St Kildans rarely ate fish, considering it too difficult and dangerous to obtain. Instead, they took hundreds of thousands of seabirds: mainly Fulmars, Puffins and Gannets. They used the birds for almost everything: food (either fresh or dried), bedding (the feathers), cooking and heating oil (from Fulmar chicks) and even slippers (whole Gannets!). They even paid their rent with feathers and skins of seabirds. No wonder anthropologists have named them the 'bird people'.

Going to pot

In ancient and medieval times, all sorts of wild birds were harvested for the pot: from Bitterns and herons to Skylarks (larks' tongues being a favourite at Roman feasts). Menus from Tudor banquets sometimes read like a birder's life list, with such unusual fare as Black-tailed Godwits, Ruff and even Dotterel being served.

Today, very few wild birds are harvested in Britain, unless you count the gamebird industry (see opposite). Nevertheless, in north-west Scotland a few individuals are still licensed to catch and kill young Gannets (known as 'guga') which are dried and sold as a local delicacy. In other parts of the world, however, this takes place on a truly industrial scale: in Australia, it has been estimated that more than half a million nesting 'mutton birds' (Short-tailed Shearwaters) are harvested every year.

In continental Europe, especially around the Mediterranean, the taking of wild birds for sport and also food is a major problem. Migrating birds in particular are targeted; shotguns and traps are a common sight on islands such as Malta and (until protests by visiting birdwatchers curtailed the practice) Cyprus. The process of using 'bird-lime' – a sticky substance which traps small birds as they perch and then prevents them from escaping – is particularly unpleasant; as are the habit of pickling songbirds such as Nightingales and the roasting of Ortolan Buntings, a particular favourite of the French.

LEFT: Puffins were once eaten by the St Kildans. OPPOSITE: The Red-legged Partridge was introduced for shooting.

Gamebirds

The definition of a gamebird is a loose one – it includes those species which are hunted for organised sport, i.e. Pheasant, partridge, grouse (members of the order *Galliformes*, and the narrowest definition of gamebirds) as well as Snipe, Woodcock, and pigeons. Although ducks and geese are also hunted, they tend to be referred to as 'wildfowl'.

Many of these species are legally defined as 'game', with special seasons when they can be hunted; the most famous of these – the grouse season – begins on August 12th, known to shooters as 'the Glorious Twelfth'. The law also requires that anyone shooting these birds holds a licence.

Shooting game, especially grouse, is now a highly lucrative industry: shooters can be charged thousands of pounds for a single day on the moor. Pheasant are specially bred for release into the wild to be subsequently shot. Indeed, several species of game, including the Pheasant itself, were originally introduced into Britain specifically for the purpose of hunting – the Red-legged or 'French' Partridge is another example.

Purists often disdain such easy targets as Pheasants, which are driven towards the guns by beaters and dogs. Instead they prefer to pit their wits against truly wild birds, with two fast-flying waders, Snipe and Woodcock, being the most sought-after and hard-to-shoot quarry.

Honeyguides

Not all human exploitation of birds is a one-way process, with little or no benefit to the birds. In Africa and Asia members of one family, the honeyguides, have developed what is essentially a symbiotic relationship between themselves and man.

Honeyguides, as their name suggests, feed on honeycomb and beeswax, which they obtain from wild bees' nests. However, if they tried to gain entry to a bees' nest on their own they would risk attack, and be driven away. So they have learned to attract a more powerful mammal, the Ratel (also known as the Honey-badger) towards the nest, watch it as it obtains entry, then fly down to feast on the remains.

In some areas, humans have observed this behaviour and learned to follow it. The bird itself appears to understand this and gives out a loud chattering call to point the human followers in the right direction. Once they have found the nest, the honeyguide will sit patiently until they have taken what they need before feeding on what is left.

Birds in folklore and culture

From predicting the weather to Greek plays and modern icons, birds have a place in mythology, art, music and literature. Early attempts to interpret bird behaviour has led to many an old wives' tale. Omens of good luck as well as bad, birds are depicted in stories and paintings in a variety of lights, often influencing how we view them today.

WEATHER LORE

From the very first time a Neolithic farmer looked up into the skies and wondered whether it would rain the next day, mankind has tried to forecast the weather. Even with today's vast supercomputers, with their unimaginable speed and capacity, meteorologists sometimes struggle to get their predictions right; so imagine how difficult it must have been for our ancestors.

Given that birds' lives are inextricably bound up with the weather, it is hardly surprising that their habits and behaviour have given rise to a wealth of folklore, which evolved as practical ways of forecasting the weather for the next day, week, month or season.

Unsettled future

Knowing if it was likely to rain, and predicting changes in the weather from settled conditions to more unsettled ones, was crucial for early farmers, as they needed to decide when to plant and harvest their crops. By observing the behaviour of certain species, especially those that feed on flying insects, they were able to make a reasonably accurate prediction of the weather, at least for the coming day or two.

OPPOSITE: Honeyguides have evolved a unique relationship with certain types of mammal. This enables both parties to benefit from the finding of bees' nests.

So there are many proverbs concerning the behaviour of swallows, in particular regarding the height at which they are feeding. High-flying swallows suggest that settled, clear conditions will remain; while if they are flying low, it means that the weather is likely to change within the next twenty-four to forty-eight hours.

Woodpeckers, too, are linked to the coming of rain; they are known as 'rain birds' in several cultures. Whether or not they really do drum or call more often when rain is on its way is a matter for debate; this may simply be a trick of acoustics, with the sound carrying further in such conditions.

Divers – known in North America as loons – are definitely known to call more frequently when wet weather is on the way. So in Shetland the Red-throated Diver is known as the 'rain goose', with its haunting call a sure precursor of wet conditions.

Rooks, too, are said to bring unsettled weather, by virtue of their 'tumbling': an acrobatic flight in which they appear to be playing together in the sky. This tends to take place in autumn during windy conditions, a fairly sure sign that rain is on the way.

The season ahead

All these beliefs, sayings and proverbs

relate to the short-term weather conditions: will it rain or be fine tomorrow? But for our ancestors, long-term forecasting was arguably even more crucial. Knowing whether this year would see a good or bad harvest, and a cold or wet winter, was absolutely critical; and without the techniques used by modern forecasters, people relied on more primitive superstition.

It has long been supposed that large numbers of geese arriving from the north in autumn, or flocks coming especially early in the season, presages a cold winter. In fact, the geese – like other species of wildfowl such as ducks and swans – are simply reacting to the local and current weather conditions. So despite a wealth of folklore related to the arrival of geese in autumn, the birds are no help in forecasting the actual weather to come. The same applies to a long series of superstitions related to the weather and bird behaviour on various days in the calendar. In short, all long-term forecasting – even using current high-tech methods, is likely to be highly inaccurate; folklore using bird behaviour to predict the distant future is likely to be even more unreliable!

BIRDS IN LITERATURE, ART AND MUSIC

From the earliest times to the present day, birds have occupied a central place in the artistic, literary and musical lives of human beings – indeed, in all aspects of high and popular culture. Some have suggested that this may be a result of their ubiquity – every culture has birds, and in most places they are the most visible and obvious of all creatures. There are also more spiritual reasons: bird song has a magical quality; we admire the beauty of their plumages and elaborate courtship displays; and perhaps most of all, we envy and admire their ability to escape the 'surly bonds of earth' – their power of flight.

Early cultures such as the Ancient Egyptians and Greeks frequently depicted birds in their ceremonial art; for example on the tombs of the Pharaohs or on friezes decorating temples (see page 170). This often related to mythology and folklore: some birds were seen as protectors; others as omens, good or bad.

Marginal mentions

In a later culture, birds often appear in the decorative margins of illuminated manuscripts of the Middle Ages, while Renaissance artists again exploited their

The Stormcock

Many birds have folk names related to their behaviour during certain kinds of weather, or to their supposed ability to predict the weather to come. Apart from those already mentioned, the 'rain bird' (various kinds of woodpecker) and 'rain goose' (Red-throated Diver), perhaps the best known is the name 'stormcock'. This has long been applied to our largest member of the thrush family, the Mistle Thrush, because of its regular habit of singing before and even during rain.

This may simply be the result of two aspects of the species' song. First, it begins to sing earlier in the year than most birds, often being in full voice as early as January. This is traditionally a time for bad weather, so the fact that the bird continues to sing through such conditions may have led to its name. Another possible reason is that unlike birds such as the Nightingale, which delivers its song hidden deep inside a bush or other foliage, Mistle Thrushes choose the highest, most prominent song post they can find, where they are likely to face the full brunt of the stormy weather.

OPPOSITE: In Shetland the Red-throated Diver is traditionally linked with bad weather and known as the 'rain goose'.

BELOW: The Mistle Thrush is sometimes known as the 'stormcock' because it often sings during wind and rain.

mythological properties, including them in allegorical paintings. Today, much of the deeper meaning of these pictures has been lost and is only accessible through careful scholarship and detective work. Thus we know that a Goldfinch in a portrait of Jesus is not merely an artistic whim, but a way of depicting the myth that the Goldfinch tried to pull the thorns from Jesus's crown to ease His suffering on the Cross, and in doing so gained the red patch on its face – supposedly from blood.

Birds with black plumage, such as crows, Ravens and even the Blackbird itself, were usually associated with evil, while other birds were also associated with particular virtues or vices: the Partridge with Truth, the peacock (not surprisingly) with Pride and Vanity, and the eagle with the Resurrection of Christ.

Owls, of course, have long symbolised wisdom, perhaps because their forward-facing appearance gives them a look more human than avian. In Japanese art, cranes are often used, again because their upright stance links them with humans – we feel closer to creatures that resemble us, however vaguely.

ABOVE: Shelley's 'blithe spirit', the Skylark is only rivalled by the Nightingale in the affections of British poets.

Birds in words

Literary references to birds are not quite as prominent as they are in art, presumably because writing lacks the visual quality that makes birds so appealing to artists. Nevertheless there are many examples of references to birds, especially in Romantic poetry of the 19th century, whose writers were primarily obsessed with birds in their symbolic capacity. Nightingales were especially popular: Keats's famous *Ode to a Nightingale* being the most famous example. The Skylark was also widely used, as in Shelley's poem *To a Skylark*.

In fact, neither Keats nor Shelley knew very much about birds, and Keats's poetry in particular is full of elementary misunderstandings about their behaviour. For ornithological accuracy in poetry, we need to look at the poems of their contemporary John Clare. Born and raised in the Northamptonshire countryside at the turn of the 19th century, Clare was a keen observer of birds and made a valuable contribution to ornithology as well as literature. His *Nightingale* is the product of both genuine knowledge and observation, and whilst the poem may not be as famous as Keats's, nor perhaps have the same literary merit, it arguably rings truer today.

Later in the 19th century the Jesuit priest Gerard Manley Hopkins was also a keen observer, who revelled in the sheer beauty of birds and the rest of God's creation. His poem *The Windhover*, comparing a hovering Kestrel to Christ, is one of the finest examples of nature and religious poetry.

It is no surprise that birds – and in particular representations of birdsong – have appeared in music throughout the centuries. Easily recognisable songs such as that of the Cuckoo, and ornate ones such as the elaborate song of the Nightingale, frequently occur: the latter as a motif in Beethoven's Symphony no. 6, also known as the *Pastoral*. Even the Beatles were not immune to the charm of birdsong, as Paul McCartney's song *Blackbird* bears witness.

BIRDS IN POPULAR CULTURE

If birds are well represented in high culture, art and literature, they are perhaps even more prevalent in our popular culture: especially in advertising, pop music and brand names. And if you delve beneath the surface there are some fascination connections between the world of pop culture and that of birds and birdwatching.

To take one example: the swan on a box of Swan Vestas matches is a direct copy of a photograph specially commissioned from the doyen of 20th-century British bird photographers, Eric Hosking. Hosking's photograph of a Montagu's Harrier was also adopted as the insignia for an RAF platoon, while the name harrier was applied to a highly manoeuvrable aircraft, the harrier jump-jet.

Raising a glass

Other links in brand names are purely coincidental: Kestrel lager perhaps suggests an uncompromising product, while Famous Grouse whisky evokes the windswept grouse moors. In the days when the Red Grouse was considered to be Britain's only endemic bird species (it is now thought to be a race of the Willow Grouse), it was used as a logo by *British Birds* magazine, whose editor then persuaded the whisky makers to act as sponsors. The most famous bird in advertising, however, is surely the Guinness Toucan – an entirely spurious but highly effective link between this comical looking bird and the sales of beer.

BELOW: The Red Grouse. has been adopted as a patriotic symbol by the makers of this eponymous brand of Scotch whisky.

Conservation and bird protection

The modern conservation movement, and in particular bird protectionists, owe their origins to the excesses of the Victorian era which placed huge demands on wildlife worldwide, especially birds.

BIRD PROTECTION SOCIETIES

At some point in the final decade of the 20th century, the one-millionth member signed up to join the biggest bird protection and conservation society not only in Britain but throughout the whole of Europe: the Royal Society for the Protection of Birds, now better known by its initials, the RSPB. In doing so, they became the latest in a long line of bird-lovers to sign up to the Society and its aims, a process that had begun more than a century earlier, in the unlikely setting of a Manchester suburb.

The RSPB was founded by a group of women who had become increasingly appalled at the flagrant exploitation of wild birds and their feathers for women's fashion. The Victorian era saw a huge rise in demand for feathers, plumes and skins, all harvested from the wild to supply what appeared to be an insatiable demand. In both Britain and North America it was a common sight to see fashionable ladies sporting a varied array of plumes from several dozen different bird species, especially on their ludicrously ornate hats.

The effect on wild bird populations is difficult to gauge, but for some species must have been devastating. We know that the Great Crested Grebe population fell to just fifty pairs by the year 1860, and by the end of the 19th century the species appeared to be doomed as a British breeding bird. Egrets were even more in demand, with species on both sides of the Atlantic being hunted for their showy white plumes.

Early society

Fortunately for the egrets and grebes, as well as for all other affected species, salvation was at hand in the unlikely form of a small group of respectable society women. Outraged at the waste and cruelty involved in the feather trade and at the indifference shown by their peers, in 1889 they gathered together to form the Society for the Protection of Birds (the Royal was added a decade or so later). So began the movement for bird protection that would come to dominate the following century and beyond.

The RSPB had its teething troubles but soon recruited like-minded souls to its cause, and by raising money through subscriptions and appeals it began to change the culture of cruelty that had so characterised Victorian attitudes towards birds. By the 1920s shooting, collecting and allied trades such as taxidermy were in steep decline, thanks partly to the efforts of writers such as W.H. Hudson and T.A. Coward, who popularised the benign and non-interventionist study of nature using binoculars and books, rather than guns.

Legal protection

Soon after the Second World War, the RSPB accomplished perhaps its greatest achievement, when with the help of other conservation bodies it persuaded the

ABOVE: A 20th-century conservation success story, the Avocet is the symbol of the Royal Society for the Protection of Birds.

government to pass the Protection of Birds Acts. These gave much greater protection than before, with almost all British species, apart from a few pests such as crows and pigeons, being protected. Indeed the Acts protected not only the birds themselves, but also their eggs, eventually putting paid to the common schoolboy pastime of egg-collecting. Nevertheless this continued to be practised illegally by a few incorrigible obsessives.

Today, the RSPB plays a central role in all aspects of bird protection and conservation, campaigning against changes in land use, bird persecution and pollution, and keeping a close watch on the likely effects of global warming on birds. The Society also manages large areas of land as nature reserves, and is now helping recreate and restore degraded habitats. It also runs many initiatives designed to increase people's interest in birds, including the running of watchpoints at key nesting sites for species such as Osprey, Peregrine and White-tailed Eagle. It is all a far cry from the origins of the movement, but is nevertheless one of the great success stories of modern conservation.

The Audubon Movement

On the other side of the Atlantic, the Audubon Societies – the North American equivalent of the RSPB – were begun at roughly the same time, and for exactly the same reason: to stop the exploitation of birds for female vanity. The movement had a rockier start than in Britain, however, perhaps because the right to shoot and keep guns is more ingrained in American culture. Americans also tend to have a more ambivalent view of wilderness and wild creatures than the British, perhaps because of the greater danger faced by the pioneers of the Wild West.

Nevertheless, the movement soon began to have widespread influence, thanks partly to the work of US President Theodore Roosevelt, who was a keen bird collector turned protector. In 1903 he set up what was probably the first official wildlife reserve in the world, in Florida; he also passed several major bird protection acts during his lifetime. Sadly, these came too late to prevent the extinction of two iconic North American species, the Passenger Pigeon and Carolina Parakeet, both of which disappeared in the early 20th century.

Birdwatching

People have always watched birds – or at least been aware of them and noted their presence. As we have already seen, the Ancient Egyptians, early Biblical scholars and the Ancient Greeks all used birds in their art or writings, and to some extent studied them and their behaviour as well. But there is a big gulf between watching birds for reasons of superstition or science and 'birdwatching' in the current sense of the word.

HOW BIRDING BEGAN

To 'go birdwatching' implies that you are taking part in an activity undertaken primarily for pleasure, as part of your leisure time – a rather modern concept. Until very recently, the lives of most human beings – apart perhaps from a few aristocrats and royals – were, in the memorable words of one Renaissance philosopher, 'poor, nasty, brutish and short'. Time was mainly taken up with work – indeed, for agricultural workers (and almost everyone did work on the land) virtually all the hours between dawn and dusk were spent working in the fields. Sundays were for rest and worship, which left no time for anything we might regard as a 'hobby'.

This began to change from the 18th century onwards, when people moved in large numbers from the countryside into the cities as a result of the Industrial Revolution sweeping Britain. A new sector of society gradually emerged: people who began to consider themselves 'middle class', and who were neither manual labourers nor titled landowners.

Nature rambles

During this period – from the late 1700s through to the end of the 19th century – leisure time began to emerge as a separate concept and, as the novels of Jane Austen and the Brontë sisters show, this time was fulfilled by all sorts of pastimes. For women, these mainly involved staying at home and arranging flowers, playing the piano or reading; but for men, going out into the open air and hunting, shooting and fishing were considered more manly activities. Both sexes went on country walks, often with a purpose such as 'botanising' – the study of wild flowers.

It was inevitable that these people should also notice birds. The practice of shooting and collecting birds as trophies was very popular, but gradually a more benign approach took hold, thanks in a great part to the writings of Gilbert White, a country vicar of the parish of Selborne in Hampshire.

White's legacy

White's famous book, *A Natural History of Selborne*, was an excellent primer in what would later be called 'nature study'. He was a close and meticulous observer, who despite not possessing any advanced optical aids – nor, indeed, any identification guides – was able to piece together specific observations into broader theories. Primarily, however, White watched birds for pleasure; thus making

OPPOSITE: The rise of the conservation movement in North America sadly came too late for the Carolina Parakeet.

ABOVE: White's Thrush, a rare vagrant to Britain from Asia, is named after the 18th-century naturalist Gilbert White.

OPPOSITE: Illustrated field guides, such as these examples, are an essential tool for birders worldwide.

him, in many people's opinion, the first real 'birdwatcher'. His chief legacy is his book, which has reputedly sold more copies than almost any work of non-fiction ever published, but he is also commemorated in the name of the largest thrush on the British list, an Asian species called White's Thrush.

FIELD GUIDES

Identifying birds is not always easy. Many species look very similar to one another, while in some species individuals can appear very different indeed, depending on whether they are an adult or young bird, male or female, or seen during or outside the breeding season.

So how do you go about identifying birds? Well, rather like playing a musical instrument or speaking a foreign language, experience and constant practice are a great help. However, even long-standing birders can sometimes be caught out by 'the one that got away', a bird that flew off without being identified.

For this reason most birders – whether they admit it or not – rely at least some of the time on a field guide. This is a small volume containing illustrations and descriptions of the birds of a particular area. As its name suggests, it is designed to be used whilst out watching birds in the field.

First in the field

Field guides are a relatively recent invention, although the Victorians had books which served a limited purpose in helping them pin down a particular identity. It is generally acknowledged that the very first proper field guide was created by the American bird artist Roger Tory Peterson. Published in 1934, *A Field Guide to the Birds* covered all the species regularly found in the eastern part of North America. But it was not the contents that singled it out from previous attempts at providing a field guide, but the revolutionary new approach. Peterson depicted birds not from an artistic point of view, but in a uniform, schematic style, in order to help birders pick out key 'field marks' and tell similar species apart. The book was an overnight success, and set the trend for guides for almost half a century afterwards.

Peterson joined forces with two British ornithologists, Guy Mountfort and Phil Hollom, to produce *A Field Guide to the Birds of Britain and Europe* in 1954. This was not the first British field guide – Richard Fitter and Richard Richardson had laid claim to that title two years earlier – but it was by far the most successful, influencing many generations of birders.

Today there are field guides to almost every part of the world, and for those whose interest in natural history stretches further than just birds, to almost every group of animals and plants. Interestingly, most still use illustrations rather than photographs, following Peterson's precept that it is better to show a stylized version of a bird which can easily be compared with similar species, than rely on the vagaries of photography, in which light, angle and the characteristics of the individual being photographed make the resulting picture less useful than the work of an artist.

TWITCHING

If there is one thing that is guaranteed to

annoy most birdwatchers (or birders, as they are now generally called) it is when journalists, broadcasters or headline-writers refer to them as 'twitchers'. Not that there is anything wrong with twitching. It is simply that twitchers are a sub-group of birders with very specific aims: the single-minded pursuit of rare birds in order to add them to their list.

Twitching has been happening ever since people travelled to see birds, although it was once called by a variety of other names. On this side of the Atlantic we had 'tally-hunters', 'tick-hunters' and 'pot-hunters'; in North America they were mainly called 'listers'. During the 1950s the word twitcher emerged and had gained widespread currency by the 1970s, when the pastime really took off.

BELOW: Cream-coloured Courser, from North Africa, is one of the most 'twitchable' vagrants to Britain. This individual appeared on Scilly in autumn 2004.

Freedom to travel

Twitching depends on three main factors. First, it requires the presence of a rare or vagrant bird, which has gone off course from its normal haunts and turned up unexpectedly at a particular location. Second, the ability to disseminate and receive information on the bird's exact whereabouts and whether or not it is still present. And third, the ability to travel rapidly and economically in order to see it.

For long periods of the 20th century the first and third factors were in place, though during the two world wars travel was considered a luxury, severely restricting the ability or inclination of people to chase after rare birds. From the 1950s onwards, as petrol rationing came to an end and the creation of the motorway system made it easier to drive long distances, twitching developed in popularity. But one problem remained: how could birders who had discovered a rarity on their local patch let people know that it was there?

World listers

There is a very special breed of twitcher known as the 'world lister', who has the time, money, and obsessive nature to spend his or her life pursuing the aim of seeing as many of the world's birds as possible. Most of these people are American; all are very rich; and all are, without doubt, slightly mad.

The problem with world listing is that at first it is easy to add dozens – even hundreds – of species on your first trip to a new location or country. However, as these listers get above the four or five thousand mark it becomes increasingly difficult. By the end they are spending indecently large sums of cash in order to try to see just a single extremely rare and elusive species; and if they fail to do so, they have wasted what for many birders would be the trip of a lifetime to some far-flung and exotic location.

Queen of the world listers, until her untimely death in 1999, was Phoebe Snetsinger. Phoebe had been diagnosed with inoperable skin cancer in her late forties, and decided to make one last trip to see birds in Australia. When the cancer went into remission, she went into overdrive; at the time of her death in Madagascar, at the age of 69, she had seen roughly 8,600 of the world's birds – at least 85 per cent of the possible total.

Heard it through the grapevine

During the 1970s, a telephone 'grapevine' system was set up, in which a network of birders contacted each other as soon as they heard the news of a rare bird. This was highly effective, but still relied on a small network of people. Then, in the late 1980s, new telecommunications technology emerged that would revolutionise twitching. Premium telephone information lines, closely followed in the 1990s by personal pagers, allowed twitchers to receive up-to-the-minute, accurate information regarding the whereabouts and presence or absence of a rare bird. The twitching revolution had begun.

Some people have gone to extraordinary lengths to see such birds, even chartering light aircraft or boats in order to reach their destination, and spending hundreds – sometimes even thousands – of pounds in doing so. The most prized species are 'firsts for Britain' – a bird that has never been recorded in the wild here before, such as the Black Lark on Anglesey in June 2003.

Today, many birders occasionally 'go twitching', especially if a rarity turns up in their local area. Others consider it a waste of time and money, and prefer to find their own birds!

BIRDING TODAY

Today, birdwatching – or birding as it is now generally known on both sides of the Atlantic, is a global pastime, attracting millions of participants. However, it remains most popular in the two nations where it really began: Britain and the USA. For some reason – perhaps our temperament, or the presence of a large, affluent middle class with plenty of spare time – the Brits and Yanks simply cannot seem to get enough of birding.

Birding is now a major leisure industry in every sense of the word. Visit the British Birdwatching Fair held at Rutland Water each August, or one of the innumerable bird events, fairs and festivals held throughout the United States, and you will see just how much the word 'industry' threatens to replace the word 'leisure'. Manufacturers and retailers of state-of-the-art optical equipment including binoculars, telescopes and digital cameras; of bird books on every possible aspect of birds and birding; of luxury guided tours to see the birds of all seven continents – including Antarctica – all reveal just how much money is involved.

BELOW: Twitchers flock each autumn to places like the Isles of Scilly in search of rare and vagrant birds.

Working with birds

On the good side, birding helps employ a lot of people (many of them birders themselves) and brings much-needed injections of cash to local economies in Britain, North America and indeed throughout the world. Birding holiday companies often employ local guides, and stay in accommodation owned by local people rather than multinational hotel chains. Birding can be done in all seasons; holiday resorts such as Cape May in New Jersey, and the Isles of Scilly, receive an influx of visitors in spring and autumn to supplement the usual summer trade.

Birding is also a useful educational tool, helping educate children and adults about the environment and nature both in the western and developing worlds – a crucial weapon in the battle to save habitats and

ABOVE: Volunteers from LIPU, the Italian League for Bird Protection, monitor migrating raptors – and help stop illegal hunting – along the Straits of Messina.

declining species. In the past, boycotts by birdwatchers have persuaded governments to ban the shooting of migratory birds – as happened on Cyprus, although not yet, unfortunately, on Malta.

Available to all

One of the great appeals of birding, of course, is that it can be done at virtually every level: from the obsessive world lister, through birders who regularly go on trips abroad or for weekend trips in their own country, to people who simply enjoy feeding birds in their back garden. It may well be the case, in fact, that someone who only practises back garden birding gets more out of it than the obsessive world lister, to whom new birds are a drug which they can never quite get enough of!

So where is birding going? Although it is certainly true that there are more active birders out in the field than ever before, and that they approach their hobby in a wider range of ways than previously, there may be a ticking time bomb waiting to go off. Many people on both sides of the Atlantic are growing aware that there are fewer and fewer young people taking up birding. This is likely to be the result of several factors: the lack of freedom youngsters have nowadays to just explore their surroundings; the huge competition from other interests and hobbies; or simply because the image of birding is just not very cool. Whatever the reason for this, unless the tide turns, within a decade or two birding may be the preserve of an older generation.

In the meantime, however, those that have discovered the joys of birding are determined to enjoy it to the full: visiting more places, meeting more people, and seeing more birds, than ever before.

GLOSSARY

Abmigration
Phenomenon whereby an individual from one species joins a flock of another and ends up migrating to a different destination with them.

Altricial
Bird that is born in an undeveloped state and remains in the nest, entirely dependent on its parents, for some time after birth. Also known as nidiculous.

Brood parasitism
The practice of laying eggs in other birds' nests, so that the young are raised by a bird that is not their parent – or, in the case of the Cuckoo, not even of their species.

Dimorphism
Difference in appearance – e.g. between male and female of same species.

Eclipse
The plumage adopted by male ducks of certain species when moulting, in which they resemble the female.

Endemic
Found only in a particular area and nowhere else in the world: usually an island, country etc.

Gliding
Energy-saving flight technique using straight, stiff wings in order to move forward without flapping.

Irruption
Irregular movement by a species in search of food, which brings it into an area where it is not found every year.

Kleptoparasitism
Piracy by one bird on another, harrying them in order to steal their food.

Lekking
An unusual form of courtship behaviour practiced by a few species, in which males display to one another in a group, with the aim of winning the attentions of a female.

Moult
The loss of old, worn feathers and their replacement with new, fresh ones.

Overshooting
Phenomenon whereby a returning migratory bird overflies its intended destination to end up in an unfamiliar area.

Passage Migrant
Species passing through an area where it neither breeds nor winters; usually in spring and/or autumn.

Passerine
Member of the largest order of birds, comprising more than half the world's species (more than 6,000); includes all songbirds.

Pelagic
Ocean-going; used of seabirds and seabird watchers!

Precocial
Bird that is able to leave the nest, swim or walk, and feed, almost as soon as it is born. Also known as nidifugous.

Predator
Bird (or other animal) that feeds on other animals, usually by killing them.

Primaries
The longest feathers on a bird's wing, used for flight.

Raptor
Member of the order Falconiformes: the day-flying birds of prey such as hawks, eagles and vultures.

Relict species
A species whose range has become fragmented, so that it only survives in a small part of the area in which it once lived.

Roosting
The action of going to a place to sleep, often with other birds.

Scavenger
Bird (or other animal) that feeds on meat taken from a dead animal.

Soaring
Flight technique used by larger, broad-winged birds to stay aloft without using too much energy.

Syrinx
Organ used uniquely by birds in order to produce song.

Territory
Area defended by a bird in order to breed, nest and raise young; occasionally also refers to area defended outside the breeding season for feeding.

Thermals
Rising currents of warm air used by large birds such as raptors in order to gain height.

Vagrant
A bird that has strayed out of its normal range.

FURTHER READING

Attracting Birds to Your Garden
Stephen Moss & David Cottridge
New Holland Publishers, 2000

Bill Oddie's Birds of Britain and Ireland
Bill Oddie
New Holland Publishers, 2004

A Bird in the Bush
Stephen Moss
Aurum Press, 2004

Birds Britannica
Mark Cocker & Richard Mabey
Chatto & Windus, 2005

The Birds of the Western Palearctic: Concise Edition
David Snow & Christopher Perrins
Oxford University Press, 1997

Birds of the World: a Checklist
James F. Clements
Ibis Publishing, 2000

Collins Bird Guide
Lars Svensson, Peter Grant, Killian Mullarney & Dan Zetterstrom
Harper Collins, 1999

Collins Field Guide to Bird Songs and Calls
Geoff Sample
Harper Collins, 1996

The Complete Garden Bird Book
Mark Golley, Stephen Moss & David Daly
New Holland Publishers, 2001

Everything you always wanted to know about birds – but were afraid to ask!
Stephen Moss
Christopher Helm, 2005

The Garden Bird Handbook
Stephen Moss
New Holland Publishers, 2003

Handbook of the Birds of the World
Josep Del Hoyo, Andrew Elliott & Jordi Sargatal
Lynx Edicions, 1994

How to Birdwatch
Stephen Moss
New Holland Publishers, 2003

The Migration Atlas
Wernham et al
Poyser, 2002

The New Encyclopaedia of Birds
Edited by Christopher Perrins
Oxford University Press, 2003

The Pocket Guide to Birds of Britain and North-West Europe
Chris Kightley, Steve Madge & Dave Nurney
Pica Press, 1998

RSPB Children's Guide to Birdwatching
David Chandler & Mike Unwin
A&C Black, 2005

RSPB Handbook of British Birds
Peter Holden & Tim Cleeves
Christopher Helm, 2002

The Secret Lives of Garden Birds
Dominic Couzens
A&C Black, 2004

The State of the Nation's Birds
Chris Mead
Whittet Books, 2000

Threatened Birds of the World
Lynx Edicions and BirdLife International, 2000

Understanding Bird Behaviour
Stephen Moss
New Holland Publishers, 2003

Where to Watch Birds in Britain
Simon Harrap & Nigel Redman
Christopher Helm, 2003

USEFUL ADDRESSES

RSPB (Royal Society for the Protection of Birds)
The Lodge, Sandy, Beds SG19 2DL
Tel: 01767 680551
www.rspb.org.uk
The RSPB is Britain's leading bird conservation organisation, with almost one million members. It runs more than 100 bird reserves up and down the country, and has a national network of members' groups. Members receive four copies of Birds magazine each year, while new members receive a gift on joining.
The junior arm of the RSPB, the Wildlife Explorers), is for members up to the age of sixteen.

BTO (British Trust for Ornithology)
The National Centre for Ornithology,
The Nunnery, Thetford, Norfolk IP24 2PU
Tel: 01842 750050
www.bto.org
The BTO offers birdwatchers the opportunity to learn more about birds by taking part in surveys such as the Garden BirdWatch or the Nest Record Scheme. BTO members also receive a bi-monthly magazine, BTO News.

The Wildlife Trusts
FREEPOST MID20441, Newark, Notts NG24 4BR
0870 0367711
www.wildlifetrusts.org
The Wildlife Trusts partnership is the UK's leading voluntary organisation covering all areas of wildlife and nature conservation, with almost 400,000 members.

WWT (Wildfowl and Wetlands Trust)
Slimbridge, Glos GL2 7BT
01453 891900
www.wwt.org.uk
The WWT is primarily dedicated to the conservation of the world's wetlands and their birds. It runs nine centres in the UK, including the Wetland Centre in Barnes, London, and the famous headquarters at Slimbridge in Gloucestershire. Members receive a quarterly magazine, Wildfowl and Wetlands, and get free entry to WWT centres.

CJ Wildbird Foods Ltd
The Rea, Upton Magna, Shrewsbury, Shropshire SY4 4UB
Tel: 0800 731 2820 (Freephone)
www.birdfood.co.uk
CJ Wildbird Foods is Britain's leading supplier of bird feeders and foodstuffs, via mail order. The company sponsors the BTO Garden BirdWatch survey, and also produce a free handbook of garden feeding, containing advice on feeding garden birds, and a catalogue of products.

Wildsounds
Dept HTWB, Cross Street, Salthouse, Norfolk NR25 7XH
Tel: 01263 741100
www.wildsounds.com
Wildsounds is Britain's leading supplier of birdsong tapes and CDs, including several on garden birds. They also stock a range of 'Teach Yourself' products, particularly useful for the beginner.

Subbuteo Natural History Books Ltd
The Rea, Upton Magna, Shrewsbury, Shropshire SY4 4UB
Tel: 0870 010 9700
sales@subbooks.demon.co.uk
Subbuteo Books provides a fast, helpful and reliable mail order service for books on birds and other aspects of natural history, including those on garden birds. Free catalogue available on request.

INDEX

Primary references are in **bold**
Page numbers in *italics* refer to illustrations

abmigration **75**
albatrosses 17, 26, 54, 58, 81
antbirds **101**
Audubon Societies **190**
Auk
Great *166*
Little 111
auks 26, 29, 112, 179
Avocet 80, 174, *189*

bathing habits **42-43**
Bee-eater
European 21, *65*, 66, 67
Little Green 171
White-throated *99*
bee-eaters 42, 47, 71, **99**, 161
biblical birds **171**, 186
bill-clattering 124
bill shapes *see* feeding
Birdlife International 163
birds of paradise 125
birds of prey 16
(*see also* individual species)
birdsong, bird calls **114-123**
dawn chorus **116-118**
duetting **122**
echolocation 23
mimicry **119**
syrinx **117**
birdwatching **191-197**
Bittern 180
Little 66
Blackbird 91, 110, 111, 113, 114, 128
Red-winged 115, 116
Blackcap 51, 78, 115, 134, 173

boobies 84, 87
Boubou, Tropical 122
Bowerbird, Satin 125
bowerbirds 125
breeding 17, 31, 58, 60, **106-139**
behaviour 137
strategy 135-136
territory **112-113**
timing **110-111**
(*see also* courtship display, mating behaviour)
British Trust for Ornithology (BTO) 9
Brush-turkey, Australian *132*
Budgerigar 178
Bunting
Cirl 160
Corn 111, *137*, 159
Ortolan 180
Reed 119
Snow 162
buntings 22, 73, 104, 111
bustards 42
buzzards 55, 68, 92

cagebirds **178-179**
Canary 151, *178*
Canary Islands 151, 178
caracaras 82, 92, 129
Carson, Rachel 158
Cassowary 18
cave paintings 170
Chaffinch 115, 117, *118*, 144
Blue 151
chats 47, 52, 53
chicks **135-136**, 137
Chiffchaff 51, 113, 173
Chough 173
Cisticola, Zitting 161
climate change 63, 152, **159-162**
(*see also* weather)

climbing *22*
cock-of-the-rocks 126
colonization **156-157**, 166
Condor, California 150, *164*
conservation and protection 188-190
contact calls 34
coots 26, 128
cormorants 26, 43, *86*
Corncrake **160**, *160*
Courser
Cream-coloured *194*
Jerdon's 167
courtship display 109, 123, 124, **125-127**
dancing 125
food-passing 126
lekking 106, 124, 126, **139**
reverse sexual dimorphism **127**
talon-grappling 126
Cowbird, Brown-headed 165
Crane
Red-crowned *124*
Sandhill *20*
Whooping 164
cranes 35, *46-47*, 55, 124, 126
Crested Coua *150*
Crossbill 61, *111*
crows 22, 68, 82, 88
Cuckoo *138*, 139, 187
Eurasian 174
Great Spotted *66*
cuckoos 23, 47, 51, 52, 106, 150, 173
Curlew, Eskimo 163
curlews 80

Darter, African *170*
Dipper
American 28

Eurasian (or White-throated) 28
dippers 28, 29, *30*
distribution, *see* range and distribution
Diver
Red-throated *28*, 183, 185
divers 26, 28
diving **27**, 27-30, 84
Dodo 18, **19**, 166
domestic cat **90-91**
domestication 176-179
Dotterel *127*, 162, 180
Dove
Collared 156, 157
Indian Spotted 156
Rock 178
Turtle 52, 110
doves 86, 105, 171
drinking **104-105**
'drumming' 123
Duck
Eider 40
Mallard 176
Mandarin 153
Muscovy 176
Ruddy 152-153
White-headed 153
ducks 26, 30, 41, 42, 63, 135, 137, 176
eclipse, *see* plumage
Dunlin 60, 80
Dunnock 114, 138, 144, 174
dust baths **43**

Eagle
African Fish *94*
Bald *14*, 86
Golden *113*
White-tailed 86, 189
eagles 15, 55, 92, 113, 126, 171
eggs 19, **131-134**
hatching **134**
incubation **132**, 137
shape 131
size 131
egg-collecting 189
Egret
Cattle 66, 67, 85, 144, 146, 156, *157*, 161
Little 161
egrets 21, 65
Egypt, Ancient 170-171, 184
Emu 18
endemic species **150**
extinction **166**

Falcon
Barbary 92
Eleonora's 53
Gyr 92
Hobby 92
Peregrine 34, 92, *93*, 94, 146, 158, 189
Red-footed 67
Saker *172*
falconry 172
falcons 92
feathers *see* plumage 24, 27
feeding 29, 31, 32, 33-34, **76-103**
feeding strategies **81-89**
foraging 79
scavenging **82**, 88, 94
specialized bills 29, 78, **80-83**, 94, 98, 100-103
stealing **86-87**
feet, specialized 28, 129
lobed 26, 29, 31
talons 94
webbed *26*
finches 22, 34, 104
Darwin's 150
(*see also* individual species)
Flamingo 82
Lesser *142-143*
flight 14
altitude 16
at sea 17
display 17
feeding on wing 17
gliding and soaring 15
hovering 15, 98
sleeping on wing 17
'V' formation 16, 35, 55
flightless birds 18, **19**, 150, 166
(*see also* individual species)
flocking behaviour **32-35**
Flycatcher
Pied 52, 65
Spotted 110
flycatchers 21, 47, 53, 104
folklore **183-187**
frigatebird 17, 43, 86, 87
Fulmar 173, 179, 180

Galapagos Islands 150
Gallinule
Allen's *70*, 71
Purple 74
gallinules 26
gamebirds 86, 135, **181**
Gannet 3, 81, 84, *87*, 112, 180
geese 26, 35, 54, 52, 63, 75, 135, 137, 176, 184
global warming, *see* climate change
Godwit
Bar-tailed 59, 65
Black-tailed 180
godwits
Goldfinch *89*, 178
Goose
Bar-headed 16
Barnacle 178
Brent 35
Canada 152
Egyptian 153
Greylag 176, 178
Pink-footed 34
Red-breasted 171

Snow *44-45*, 75
Swan 176
White-fronted 75
Goldcrest 34, 69
Goldfinch 115, 155, 186
Gould, John 178
Grebe
Clark's 29
Great Crested 29, *41*, 126, 127, 131, 188
Least 29
Little *29*
Pied-billed *74*
grebes 26, 28, **29**, 72, 128, 132
Greece, Ancient 171, 184
Greenfinch 115
Greenshank 52
Grosbeak, Evening *11*
Grouse
Black 139
Red 187
Willow 187
grouse 135, 181
Guillemot 64, *112*, 128, 131
guineafowl 176
Gull
Ivory 71
Lesser Black-backed 39
Mediterranean 161
Ring-billed 72
Ross's 71
gulls 22, 26, 33, 41, 42, 79, *81*, *88*, 89, 136

Harrier
Hen 94
Marsh 94
Montagu's 94, 174, *175*, 187
harriers 92, 126
harvesting for food **179-181**
hawks 55, 59, 92
hearing 96
Heron
Black 83
Grey 69
Night 66
Squacco 66
herons 21, 36, 65, 83, 91, 180
Hobby 92
honeyguide **182**, *182*
Hoopoe 21, 66, 67
hopping 22
hornbills **130**, *130*
Hummingbird
Sword-billed 100
hummingbirds *12-13*, 15, *79*, *98*, **98-100**
Huxley, Julian 126-127

Ibis
Glossy 72
Sacred 171
ibises *16*
introductions **152-156**
Italian League for Bird Protection (LIPU) 197

jacana **25**
Junglefowl, Red (domestic chicken) 131, 176, *177*
juveniles 51, 52, 60

Kestrel 15, 84, 187
Mauritius 164
Kingfisher
Belted 102
Common 102
Dwarf 102
Pigmy 102
Malachite 102
Pied 84, 103
Woodland 103
kingfishers *11*, *68*, 69, 83, 84, *102*, **102-103**,131, 135
Kite
Black 94, *161*
Mississippi 67
Pariah 94
Red 94
Swallow-tailed 67
Yellow-billed 94
kites 69, 82, 92
Kittiwake 87, 173, 179
kiwis 19, 131
Knot 37, 65, 111
kookaburra 103

Lapwing *62-63*, 63, 80. 145, 159
Lark
Black 195
Shore (or Horned) *148*
Skylark *43*, 64, 159, 180, 186
larks 42
lekking (*see* courtship display)
'lily-trotter' *see* jacana
Linnet 145, 159, 178
literary references **186-187**
Lyrebird, Superb *120*

Macaw
Blue and Yellow *2*
Spix's *162*, 163, 179
magpies 134
Manakin, White-bearded 124, 137
Martin
House 110, 128-129
Sand 51, 128, 130
martins 47, *50*, 52, 54, 55, 105
mating behaviour 112-113
(*see also* courtship display)
migrants 31, 32
migration 15, 16, **44-61**
abmigration **75**
altitude **54**
altitudinal **64**
drift **66**
falls 66
irruption **60**

leapfrog **59**, 60
loop 60
migratory flocks 35, 55
navigation and orientation **48-49**
north-south **47-48**
overshoots **65-67**
ship assistance **74**
strategies 48, **53-55**
tracking 54
vagrancy 63, 66-67, **70-75**
moult *see* plumage
movement **62-64**
Murrelet, Ancient 71
Myna, Common 156
mynas 120

naming **173-175**
nesting 17, 29, 67, **128-130**
brood parasitism **138-139**
(*see also* eggs, hatching, chicks)
nighthawks 102
Nightingale *115*, 160, 180, 176, 187
nightjars 102, 123
night vision 95
nocturnal species 95-96, 102
Nuthatch
Red-breasted 74
nuthatches 24, 34

Oilbird 102, *123*
orioles 47
Osprey 79, 84, *85*, 86, 92, 146, *147*, 189
Ostrich 18, *18*, 19, 22-23, 42, 131, *134*, 178
Owl
Barn *15*, *76-77*, 95, *96*, 146
Barred 98
Burrowing *129*
Eastern Screech 98
Great Grey *11*, 96
Great Horned 98
Little 98, 153
Northern Hawk 96
Pharaoh Eagle 96
Scops 98
Snowy *61*, 97, 162
Tawny *16*, 68
owls 79, 91, 95-98, 131, 186
oxpeckers *85*
Oystercatcher *42*

Parakeet
Carolina *190*
Rose-ringed *6*, 153, *154*, 178
parakeets **154**, 155
Parrot, Burrowing 129
parrots 79, 86, 166, 179
Partridge
Grey 145, 153, *158*, 159
Red-legged 153, *181*
partridges 135, 181, 186
Peafowl, Indian (peacock) *108-109*, 109, 186
pelagic species 58, 81, 104
pelicans 15, 26, 55, *78-79*
Penguin
Emperor *8-9*, 38, *133*, **133**
Galapagos 38
King *140-141*
Magellanic 129
penguins **27**, 29
pests **34**
pesticides 158
Petrel
Leach's 174
petrels 33, 54, 81
Phalarope
Grey (or Red) *31*, 127
Red-necked 31, 127, 162
Wilson's *73*, 127
phalaropes 26, 31
pheasant 135, *152*, 181
Pigeon
Passenger 144, 190
Wood 119
pigeons 86, **104**, 105, 166, 178, 181
laurel 151
Pipit
Meadow 64, 138, 144
Water 57
pipits 22
Pitta
Gurney's 164
Plover
Pacific Golden 59
Ringed *60*, *136*
plumage 24, 27, 29, 31, 39, 52, 123, 125
breeding 41
courtship display 109
eclipse **40**
moulting **39-41**
warmth 38, 39
waterproofing 26, 30, 39, 43
potoos 102
predators 18, 34, 53, 59, 68, **90-94**, 134, 135
prey 23, **90-94**, 102-103, 123
Protection of Birds Acts 188-189
Ptarmigan 41, *64*, 92, 162
Puffin 130, 179, *180*

Quail 119
Mountain 64
Quelea, Red-billed *33*, 144

Rail
Invisible 167
Inaccessible Island Flightless 150
rails 26, 166
range and distribution **140-162**
raptors 41, 47, 55, 79, **91-94**, 126

Raven 171
Fan-tailed 171
Razorbill 64, 128
Redhead 72
Redshank, Spotted 52
Redstart 173
Redwing 63
relict species 148
rheas 18, 19, 22
Roadrunner
Greater 23
Lesser 23
roadrunner *22*, 23
Robin
European *7*, 113, 114, *115*, 138, 144, 178
robin-chats 122
robins 22, 84, 86, 104, 128
Roller 71
roosting **36-38**
Royal Society for the Protection of Birds (RSPB) 9, 188-189
Ruff 138, 180
running 22

Sanderling 22
sandgrouse **105**
Sandpiper
Broad-billed 57
Curlew 56, 57
Green 52
Spoon-billed 82
Spotted 73
Sapsucker
Yellow-bellied 74
seabirds 16, 26, 33, 48, 60, 104, 113, 130
(*see also individual species*)
Serin 161
shags 26, 112
Shearwater
Great **61**
Manx 179
Short-tailed 180
Sooty **61**
shearwaters 17, 26, 33, 54, 58, **61**, 81, *179*
shooting 181
Shoveler 82, 173
Shrike
Red-backed 56, 66, *97*, 174
Woodchat 66
shrikes **97**, 122
Skimmer *103*
African **103**
Black **103**
Skua
Great *87*
Pomarine 60
South Polar 71
skuas 33, 81, 86, 87, 129
sleeping **36-38**
Smew 63
Snipe 80, **123-124**, 181
Common *7*
Great 139
Snowcock, Himalayan 154
Solitaire **19**
songbirds 16, 39, 52, 137
(*see also* individual species)
Sparrow
House 84, 144, 146, 153, *155*
Tree 84, *159*
Sparrowhawk 4, 68, **90-91**, 139
sparrows 22, 42, 73, 104
'spinning' 31
Spoonbills 161
St Kilda 180
Starling, European 121, 144, 146, 148, *153*, 155
starlings 22, 34, 37, *38*, 120
Stilt, Black-winged 65, 67
stints 80
Stork
Marabou 33, 82
White *6*, 84, *106-107*, 124, *125*
storks 15, 47, 55, 171
Storm-petrel, Wilson's 144
Stormcock, *see* Mistle Thrush
Swallow
Barn 48, 51, 59, 128, 174
Tree *52*
swallows *6*, 47, *50*, 52, 54, 55, 105, 110, 183
Swan
Bewick's 174
Mute 131, 137
swans 26, 54, 135, 187
Swift
Alpine 66
Common *17*
swifts 16, 47, 52, 55
swimming **26-31**,
Takahe 167
Teal *40*
Tern
Aleutian 71
Arctic 48, *58*, 146
Caspian 148
Gull-billed 148
Sandwich 51
terns 17, 26, 33, 43, *49*, 87
Thrush
Mistle 100, 114, *185*
Song 110, 114, 144, 155
White's *192*
thrushes 22, 47, 52, 59, 73
Tit
Blue 78, 113, 115
Great 115, 118
Long-tailed 69, 128, 136
Marsh 173
Penduline *128*
Willow 173
tits 34, 39, 79, 128
toucan *100*
Turaco
Ross's 131

Treecreeper *24*, 34, 69, 173
tropicbirds 87
Turkey 176

Twite 148

vagrants 21
vireos 59, 72
Vulture
Black 82
Griffon 82
Rüppell's Griffon 16
Turkey 82
vultures 79, *82*, 92

waders 16, 22, 34, 26, 37, 42, 47, 48, 51, 52, 54, 60, 72, 79, 135
(*see also* individual species)
Wagtail, Pied 118
wagtails 22
Wallcreeper 24
Warbler
Aquatic 56
Cetti's 161
Dartford *69*, 160
Grasshopper 119
Great Reed 161
Icterine 66
Kirtland's 164, **165**
Marsh 52, **121**
Melodious 161
Pallas's 74, *75*
Reed 52, *138*
Rüppell's 71
Sedge 52
Willow *7*, 52, 65, 115, 173
Yellow 52, *53*
Yellow-browed 75
Wood 59, 72, 73
warblers 21, 42, 47, 52, 104, 178
Waterthrush, Louisiana 52
waterfowl 42, 63
(*see also* individual species)
Waxwing 60, 100
Cedar *72*
weather conditions 20, **62-69**
jet-stream 73
thermal air currents 15-16, 55
wind 21, 59
weaver birds 128
Wheatear, Northern 51, 59
wheatears 51, 173
Whimbrel 52
Whipbird, Eastern *122*
White, Gilbert 172, 191
Whitethroat 52, 65, 100, 173
Lesser 55
whydahs 109
widowbirds 109
wildfowl 16, 22, 36, 111, 181
(*see also* individual species)
Wildfowl Trust 152
Willet 52
Woodcock 181
Woodpecker
Ground 24
Ivory-billed *167*
woodpeckers *22*, 23, 24, 124, 128, 135, 173, 183
Wren 69, 114, 144, *145*, **151**
Wrybill 82

Yellowhammer 111, 118, *119*, 145, 159
Yellowlegs, Lesser 73

ACKNOWLEDGEMENTS

With thanks to everyone at New Holland involved in the production and promotion of this book, especially Steffanie Brown, Liz Dittner, Charlotte Judet, Adam Morris, Alan Marshall, James Parry, Victoria Scales, Yvonne Thynne and, as ever, the fabulous Jo Hemmings. Also a big thanks to Chris Harbard, whose expert eyes were cast over the original text, and whose vast knowledge of birds helped improve it beyond measure.

And, as always, thanks to my lovely wife Suzanne and the tribe of Moss offspring: David and James (who as teenagers pretend not to be interested in birds), and Charlie, George and Daisy, who are still young enough to show enthusiasm for the robins, goldfinches and parakeets they see from our back window!

PICTURE CREDITS

A N T Photo Library/NHPA page 122; B & C Alexander/NHPA page 133; Thomas Arndt/NHPA page 163; Anthony Bannister/NHPA pages 49, 50; George Bernard/NHPA pages 19, 144, 166, 167, 190; Alan Barnes page 7 (top); Jordi Bas Casas/NHPA pages 92, 149, 175; Bruce Beehler/NHPA page 192; Joe Blossom/NHPA pages 80, 179; Andrea Bonetti/NHPA pages 66, 86; Simon Booth/NHPA page 151; Mark Bowler/NHPA page 164; John & Sue Buckingham/NHPA page 114; Laurie Campbell/NHPA pages 7 (middle), 30, 112, 131, 152, 155, 156, 159, 160, 185; Robin Chittenden www.harlequinpictures.co.uk page 40; Bill Coster/NHPA pages 21, 25, 37, 52, 128, 139; N R Coulton/NHPA page 81; Stephen Dalton/NHPA pages 2, 7 (bottom), 12; 14; 15, 36, 77, 91, 95, 102; Manfred Danegger/NHPA pages 6 (middle), 90, 107, 118, 126, 158, 189; Nigel J Dennis/NHPA pages 78, 94, 99, 150, 182; Guy Edwardes/NHPA pages 60, 113; Robert Erwin/NHPA page 72; *The Famous Grouse* page 187; Melvin Grey/NHPA pages 65, 110; Hugh Harrop page 87; Martin Harvey/NHPA pages 6 (top), 85, 134, 143; Brian Hawkes/NHPA page 123; Paal Hermansen/NHPA pages 54, 184; Daniel Heuclin/NHPA page 147; Hellio & Van Ingen/NHPA pages 41, 61, 111; Ernie Janes/NHPA pages 62, 82, 89, 104; R & D Keller/NHPA page 16; Rich Kirchner/NHPA page 73; T Kitchin & V Hurst/NHPA page 88; Stephen Krasemann/NHPA pages 11 (bottom), 53, 129, 168; Gerard Lacz/NHPA page 178; Mike Lane/NHPA pages 24, 29, 42, 121, 132, 136; Michael Leach/NHPA page 171; Jean-Marie Le Moigne/NHPA pages 28, 137, 157; Michael McKee page 194; James Morris/Axiom page 170; William Paton/NHPA page 26; Daniele Pellegrini/Airone page 197; Jari Peltomaki/NHPA page 11 (top right); Pierre Pettit/NHPA pages 46, 55; P & B Pickford/NHPA page 70; Peter Pickford/NHPA page 83; Rod Planck/NHPA pages 39, 141; Christophe Ratier/NHPA page 18; Steve Robinson/NHPA page 108; Andy Rouse/NHPA pages 22, 85 (top), 100, 103, 125, 181; Robert Royse page 165; Kevin Schafer/NHPA pages 8, 27, 98, 101; John Shaw/NHPA pages 44, 58, 74; Eric Soder/NHPA pages 17, 68; Morten Strange/NHPA pages 56, 131; Roger Tidman/NHPA pages 59, 75, 124, 135, 138, 161, 196; Ann & Steve Toon/NHPA pages 32, 105, 177; James Warwick/NHPA pages 6 (bottom), 38; Dave Watts/NHPA pages 97, 120; Alan Williams/NHPA pages 5, 11 (top left), 23, 31, 35, 43, 64, 67, 69, 71, 96, 115, 116, 119, 127, 145, 148, 153, 154, 162, 180, 186.